EDITION architektur.aktuell 04

Matthias Boeckl (Ed.)

StadtUmbau | UrbanConversion

Recent International Examples

SpringerWienNewYork

Dr. M. Boeckl
Springer-Verlag Wien New York
Sachsenplatz 4-6
A-1201 Wien

Printed in Austria

Redaktion | Editorial Team: Andrea Nussbaum, Gerald A. Rödler
Übersetzung | Translations: JLQJ translations + Peter Waugh
Grafik | Graphic Design: A|H Haller, Wien
Druck | Printing: Grasl Druck & Neue Medien, A-2540 Bad Vöslau

Gedruckt auf säurefreiem, chlorfrei gebleichtem Papier – TCF |
Printed on acid-free and chlorine-free bleached paper

SPIN: 10886903

Bibliografische Information Der Deutschen Bibliothek
Die Deutsche Bibliothek verzeichnet diese Publikation in der
Deutschen Nationalbibliografie; detaillierte bibliografische Daten sind im Internet über <http://dnb.ddb.de> abrufbar.

Mit zahlreichen farbigen Abbildungen | With numerous coloured figures

ISBN 3-211-83850-3 Springer-Verlag Wien New York

INHALT | CONTENTS

Sao Paulo
gesehen von | seen by Aglaia Konrad

Stadtumbau – eine globale Notwendigkeit
Urban Conversion – A Global Necessity

Dieses Buch bietet einen Beitrag zur städtebaulich-architektonischen Abteilung der Globalisierungsdebatte. Da heute weltweit die meisten Menschen in Städten leben, ist das nicht ihr unbedeutendster Aspekt. Unter dem Schlagwort „Stadtumbau" werden heute nahezu alle Strategien zusammengefasst, die einem vorwiegend von kurzsichtigen Gewinnerwartungen statt von langfristigen Entwicklungskonzepten gesteuerten Städtebau entgegenwirken sollen. Wie Harald Bodenschatz in seinem Essay im vorliegenden Band darlegt, zählen dazu insbesondere der Umbau von Stadtzentren, die Konversion der zahlreichen von der Industriegesellschaft zurückgelassenen Brachflächen sowie die Nachbesserung monofunktionaler Stadtgebiete der gleichen Ära. Obwohl diese Debatte bis dato vorwiegend in Europa und Nordamerika handelt, sind ihre Strukturen zweifellos auch auf Städte Asiens, Afrikas und Südamerikas anwendbar. In einigen dieser Weltregionen vollzieht sich der grundlegende Wandel der Städte noch um einiges rascher als in Europa, das für den Übergang von der alten Administrations-, Residenz- und Handelsstadt über das problematische Zwischenstadium der industrialisierten Metropolis bis zur postindustriellen Produktionsstätte von Wissen und Lebensqualität immerhin zweihundert Jahre benötigte. Dieser Prozess wird andernorts innerhalb weniger Jahrzehnte abgespult, bleibt aber gerade in der „dritten" Welt oft in der negativsten Entwicklungsphase mit Umweltzerstörung und sozialem Niedergang stecken.
Schon wegen dieser weitgehenden Verquickung der Schicksals der Städte mit der weltweiten Ausbreitung der Marktwirtschaft und ihren globalen Playern sind die Grundprobleme der Globalisierungs- und der Stadtdebatte im Grunde identisch: Gelingt es, neben Errungenschaften der Industrialisierung wie Wohlstand, Versorgungssicherheit und Lebensqualität auch Verhinderungsstrategien ihrer katastrophalen Nebenwirkungen wie Desintegration, Zerstörung traditioneller Lebenswelten und Migrationsdruck global zu verbreiten? Die begleitende soziale Kontrolle einer wirtschaftsliberalen Entwicklung ist heute wertvolles politisches Know-How, das im permanenten weltweiten Datenaustausch ständig optimiert wird. Entsprechende Techniken der Stadtplanung, welche die Tendenzen zu Ghettoisierung, Gentrifikation und suburbaner Wucherung mit all ihrem sozialen Sprengstoff abmildern können, sind bekannt – aber zweifellos sind es auch die gegebenen geopolitischen Strukturen, die ihre Anwendung nicht eben befördern. Zahlreiche sinnvolle Stadtprojekte etwa in den Megalopolen Südamerikas scheitern nicht zuletzt an den politischen Verhältnissen in den betroffenen Ländern, deren nachhaltige Reform den „Großen Acht" nicht immer ins globale Machtkonzept passt.

This book offers a contribution to the segment of the globalisation debate involving architecture and urban planning. As nowadays throughout the world the majority of people live in towns and cities, this is by no means the least important aspect of this debate. Under the slogan 'urban conversion' almost all current strategies intended to combat the kind of urban design that is based primarily on short-sighted profit making instead of long-term development concepts, are presented. As Harald Bodenschatz explains in his essay in this volume, these include in particular the redesign of city centres, the conversion of numerous brownfield sites left behind by industry as well as the improvement of mononuclear urban districts dating from the same era. Although up to the present this debate has been centred for the most part in Europe and North America its structures are doubtless applicable to the cities of Asia, Africa and South America.
In several of these regions of the world the fundamental transformation of cities is taking place at a considerably faster pace than in Europe which, it should be said, took two hundred years to complete the transformation from the old administrative, court and commercial city via the problematic intermediate state of the industrialised metropolis to the post-industrial places for the production of knowledge and a quality of life. The same process is completed elsewhere within a few decades but, particularly in the so-called 'third world', it often remains stuck at the most negative development phase, involving the destruction of the environment and social decline. Alone on account of the close links between the destiny of the cities and the world-wide spread of the market economy with its global players, the basic issues in the globalisation debate are essentially identical with those raised in the discussion on cities. Can it prove possible, alongside the achievements of the industrialisation process such as prosperity, guaranteed services and quality of life, to develop globally strategies that prevent the catastrophic side effects of these benefits such as disintegration, the destruction of traditional worlds and pressure to migrate? The constant social control of liberal economic development today represents valuable know-how that is constantly being optimised in the permanent world-wide exchange of data. The related techniques of town planning that can help alleviate the problems represented by the formation of ghettos, gentrification and the uncontrolled growth of suburbia and all the socially explosive potential of these tendencies, are already familiar, but there can be little doubt that existing geopolitical structures often do little to support their application. Numerous important urban projects such as those in the mega-cities of South America fail due, not least of all, to the political situation in the respective countries, and the reform of such political constellations is not always in the interest of the global power strategies of the 'Big Eight'.

Amsterdam-Bijlmemeer
gesehen von | seen by Aglaia Konrad

Das vorliegende Buch zeigt eine akzentuierte Auswahl des Instrumentariums des Stadtumbaus. Von der Transformation ganzer Stadtbezirke – wie der South Bank in London – über intelligente Rettungsstrategien bedrohter Kleinstädte – hier am Beispiel des niederösterreichischen Bezirksstadt Waidhofen an der Ybbs – bis hin zu einzelnen Bauinterventionen nach dem Muster der hier präsentierten beiden Kulturbauten in Manhattan reicht das Spektrum aktueller Möglichkeiten. Der eingangs zitierte Beitrag von Harald Bodenschatz liefert diesen sehr unterschiedlichen Projekten das theoretische Fundament. Die nächste Phase der globalen Stadtdebatte wird jedoch die Integration von Erfahrungen der extremsten Stadtkrisen bringen müssen, wie wir sie derzeit in Bagdad, Caracas, Sao Paolo, Johannesburg und Shanghai erleben. Denn die im vorliegenden Band vorgestellten Stadtprojekte handeln in friedlichen und demokratischen Weltgegenden, während die weltweite urbane Bevölkerungsmehrheit unter wesentlich ungünstigeren Bedingungen Wege aus der Stadtapokalypse finden muss. Wenn auch das Ziel dieser langwierigen, teilweise noch nicht einmal begonnenen Stadtsanierungen in ernsthaft notleidenden Agglomerationen mit dem „europäischen" Modell weitgehend außer Streit zu stehen scheint, herrscht über die gangbaren Wege dorthin regelrechte Ratlosigkeit. Die Stadtumbaustrategien der westlichen Welt sind wegen oftmals gänzlich anderer Stadtfunktionsweisen in Schwellen- und Entwicklungsländern wohl nicht immer die richtigen. Aus den negativen Auswirkungen der ersten Globalisierungsära, nämlich jener der Industrie, sollten wir gelernt haben, dass die stereotype und normative Anwendung von Technologie ohne Rücksicht auf lokale Verhältnisse das Gegenteil von Entwicklung bewirken kann: Nämlich den Abstieg in Armut, Chaos und Anarchie. Gefragt wäre daher die Entwicklung anpassungsfähiger Planungsinstrumente des Stadtumbaus, die möglichst unabhängig von Machtverhältnissen, aber dennoch auf Basis einer sozial orientierten Warenwirtschaft funktionieren. Also nichts weniger als die Quadratur des Kreises.

Matthias Boeckl

This book presents an accentuated selection of the instruments used in urban re-planning. The spectrum of current possibilities extends from the transformation of entire urban districts – such as the South Bank in London – to intelligent rescue strategies for threatened small towns – illustrated here by the example of the Lower Austrian regional town of Waidhofen an der Ybbs – to individual building interventions according to the pattern of the two cultural buildings in Manhattan presented here. The contribution from Harald Bodenschatz referred to above offers the theoretical basis for the very different projects. The next phase of the urban debate on the city will have to include the experiences of the most extreme urban crises such as those presently existing in Baghdad, Caracas, Sao Paolo, Johannesburg and Shanghai. The urban projects presented in this volume are all located in peaceful and democratic regions of the world whereas the majority of the world population must find its way out of the urban apocalypse under far less favourable conditions. Although the goal of this lengthy process of urban renewal (which in places has not even started) that is based on 'European' models appears largely uncontested in the urban agglomerations that at present suffer serious problems, there still prevails widespread uncertainty about the paths leading to this goal. On account of the often very different ways cities function in second and third world countries, the urban re-planning strategies of the western world are not always the correct ones. From the negative consequences of the first era of globalisation, i.e. the age of industrialisation, we ought by now to have learnt that the stereotypical and normative application of technology without taking into account local conditions can produce the opposite of development, that is to say a decline into poverty chaos and anarchy. What is needed therefore is the development of a set of planning instruments for urban reconstruction which functions as independently as possible of the prevailing balance of power but still on the basis of socially oriented market economy. That is to say: nothing less than squaring the circle.

Matthias Boeckl

Stadtumbau – Begriffe und Perspektiven
Urban Conversion, Concepts and Perspectives

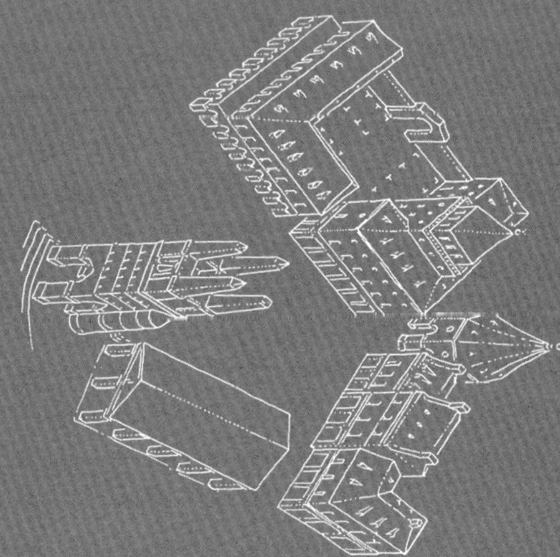

Stadtumbau – Begriffe und Perspektiven
Urban Conversion, Concepts and Perspectives

Text Harald Bodenschatz

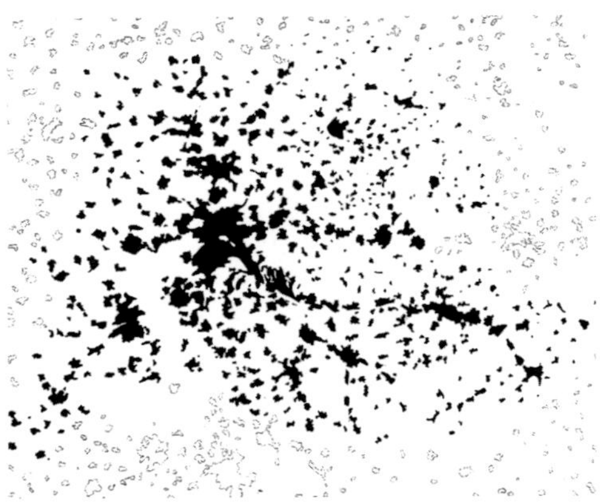

Th. Sieverts: Stuttgart 1950, 1995

Die Fachwelt ist sich einig: Wir erleben zur Zeit, so heißt es, einen „tief greifenden Wandel" – weg von der Industriegesellschaft, weg von der Familie, weg von der Alterspyramide, weg von einer Gesellschaft mit stabilen Arbeitsverhältnissen, weg von einer Polarisierung in „Ost" und „West", weg von einer Gesellschaft, die durch lokale Ereignisse geprägt wird. Wo die Reise aber hingeht, ist viel weniger deutlich: Postindustrialisierung, mehr und mehr Singles, Überalterung, Wissensgesellschaft, Globalisierung – diese beliebten Begriffe weisen in die Zukunft, ohne sie im Detail klären zu können. Was heißt das für unsere Städte? Wie und wo werden sie sich ändern müssen?

The specialist world is unanimous: we are experiencing a time, it is said, of 'profound transformation' – moving away from the industrial society, away from the family, away from the age pyramid, away from a society with stable working conditions, away from a polarisation of 'east' and 'west', away from a society characterised by local events. However, it is far less clear where the journey is actually leading us: post-industrialisation, more and more single people, over-ageing, a society of knowledge, globalisation – these popular terms indicate the future without being able to clarify it in detail.
What does it mean for our cities?
How and where will they have to change?

Halte à la renovation! Verblassender Protest gegen Kahlschlagsanierung in Paris, 1983
Halte à la renovation! Waning protest against tabula rasa demolition in Paris, 1983

Stadtumbau als Antwort auf den gesellschaftlichen Wandel

Die Antwort zumindest der deutschen Urbanisten auf den weithin unberechenbaren gesellschaftlichen Wandel heißt Stadtumbau. Ein Begriff, der vor wenigen Jahren – außer im Zusammenhang mit dem „ökologischen Stadtumbau“ – noch keine besondere Bedeutung hatte, heute aber in aller Munde ist. Eine erstaunliche Karriere! Beflügelt wurde der Begriff durch ein staatliches Förderprogramm gleichen Namens: „Stadtumbau Ost“ (seit 2002) und – weil es so gut ankommt – nun auch „Stadtumbau West“. Das Leitbild für den Stadtumbau ist – wie alle städtebaulichen Leitbilder – sehr schillernd: Ziel ist die „Rettung“ oder die (mehr oder minder „kritische“) Rekonstruktion der traditionellen Stadt, der kompakten Stadt, bzw., wie gerne in Deutschland gesagt wird, der „europäischen Stadt“. Die traditionelle Stadt ist weniger ein analytischer Begriff als ein städtebauliches Programm. Ihr werden Merkmale wie eine relativ hohe bauliche Dichte, ein vernetztes System öffentlicher Räume, eine soziale, funktionale und architektonische Mischung sowie eine räumliche Hierarchie mit einem Zentrum als Höhepunkt zugeschrieben. Die konkrete europäische Stadt ist in diesem Sinne die materielle Interpretation ihrer jeweils besonderen Geschichte, die es zu erhalten, an neue Anforderungen anzupassen bzw. zu reproduzieren gilt. Die Beschwörung der traditionellen Stadt ist zugleich eine Absage an die Stadt der „Nachkriegsmoderne” und an die „amerikanische Stadt”.
Was verbirgt sich aber praktisch hinter dem Begriff „Stadtumbau“? Damit sind heute folgende drei große Aktionsfelder gemeint: erstens der Umbau der Zentren, lange Zeit auch als „Revitalisierung“ bezeichnet; zweitens der Umbau von brach gefallenen, nicht mehr genutzten Flächen, gerne auch „Konversion“ genannt; sowie drittens der Umbau der großen, monofunktionalen Siedlungen des Sozialen Wohnungsbaus, früher als „Nachbesserung“ bekannt.

Urban Redevelopment As A Response To Social Change

The response at least of German urbanists to this still unpredictable social change is called ‘urban redevelopment’. A term which, only a few years ago – except when used in the context of ‘ecological urban redevelopment’ – still did not have any particular significance, but which is today on everybody’s lips. An astonishing career! The term was inspired by a state subsidy programme of the same name: ‘Urban Redevelopment East’ (running since 2002), which – because it was so well received – has now also been followed by ‘Urban Redevelopment West’. The leitmotif for urban redevelopment is absolutely scintillating – as are all the leitmotifs for urban development: the aim is to save or (more or less critically) reconstruct the traditional city, the compact city or, as people in Germany like to call it, the European city. The traditional city is less of an analytical term than an urban development programme. To it are attributed such characteristics as a relatively high density of buildings, a system of networked public spaces, a social, functional and architectural mix, as well as a spatial hierarchy with a city centre as the highlight. The actual European city is in this sense the material interpretation of its respective unique history, which it is important to preserve, adapt to new requirements or to reproduce. The invocation of the traditional city is at the same time a denial of the city of post-war modernism and the American city.
Yet what lies concealed behind the term ‘urban redevelopment’ in practice? Today, the following three large-scale fields of action are intended: firstly, the redevelopment of the centre, for a long time also known as ‘rejuvenation’; secondly, the redevelopment of areas that are not being used and have become wasteland, popularly also known as ‘conversion’; and thirdly, the redevelopment of large-scale, monofunctional council housing estates which was formerly known as ‘adjustment’.

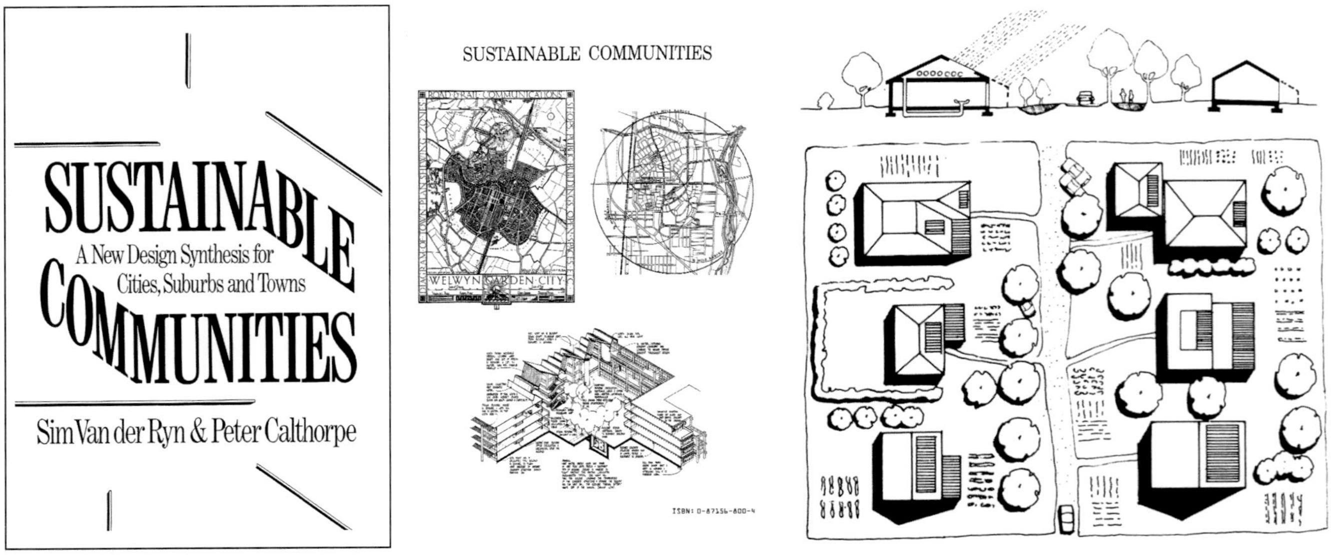

Peter Calthrope, Sustainable Communities, 1986

MuseumsQuartier Wien: Zentrumsumbau | redevelopment of a Center

Krakau: Modellstadt der erhaltenden Stadterneuerung seit den 1970er Jahren: Straßenszene 1985
Cracow: model city of preservationist urban renewal since the 1970s: street scene, 1985

Im Mittelpunkt des öffentlichen Interesses steht der Umbau der Zentren der großen Städte. Wohin dieser gehen wird, deutet sich bereits an: Das Großstadtzentrum der Zukunft wird das Zentrum einer suburbanisierten Stadtregion sein. Aber nicht als ein Fragment neben anderen, sondern als Zentrum neuer Art, in dem sich die Suburbaniten zuhause fühlen. Dort finden sie einzigartige Gebäude, die eine wenig aufregend gestaltete suburbane Landschaft dringend benötigt, aber auch ein ausgeklügeltes Angebot an Hochkultur, Unterhaltung und seltenen Waren. Das neue Zentrum ist ein Ort konstruierter Geschichte, auf die eine geschichtslose suburbane Landschaft angewiesen ist, ein Ziel des Tourismus, und zwar nicht nur des weltweit wachsenden Ferntourismus, sondern vor allem auch des suburbanen Lokaltourismus. Im neuen Zentrum ballen sich ausgewählte Institutionen der Produktion von Wissen, der Entscheidung und Kreativität. Und es gibt sogar Platz und Heimat für Leute, die dort gerne wohnen, auch alte Menschen.

Damit ist zugleich das städtebauliche Programm des Zentrumsumbaus umrissen: Der öffentliche Raum wird für Fußgänger wieder gewonnen, verschönert bzw. neu geschaffen. Spektakuläre Entertainment- und Kulturkomplexe werden neu gebaut oder in historische Gebäude implantiert. Die Nutzungen werden besser gemischt, vor allem wird auch in zentraler Lage attraktiver Wohnraum für Besserverdienende geschaffen. Das Zentrum brummt rund um die Uhr („24-Stunden-Stadt"), die Stadt orientiert sich – wo immer möglich – zum Wasser, und sei es nur zu einem brackigen, nicht immer wohlriechenden Kanal. Das städtische Grün wird erweitert und qualifiziert. Schließlich wird das städtebauliche Erbe sorgfältig gepflegt und mit historistischen Rekonstruktionen angereichert. Dazu kommen Aufsehen erregende Neubauten, die bierernst oder fröhlich Zukunftsfähigkeit symbolisieren.

Das zweite Top-Thema des Stadtumbaus betrifft die „Konversion". In, am Rande und außerhalb der Innenstädte sind in den letzten Jahrzehnten riesige Gebiete brach gefallen. Dabei handelt es sich um ehemalige Bahn-, Industrie- und Gewerbe-, Hafen-, Militär-, Messe- und Flughafenflächen. Wie können diese einer neuen Nutzung zugeführt werden? Durch eine Ausweitung der traditionellen Stadt oder durch gänzlich neue Strukturen?

Vor allem in Deutschland hat Stadtumbau seit kurzem eine dritte Bedeutung erhalten: die Anpassung der großen Wohnquartiere des industrialisierten, sozialen Massenwohnungsbaus an veränderte Bedarfe. Der „Stadtumbau Ost" dient in erster Linie dem Abbruch von „Plattenbauten", denen die Bewohner abhanden gekommen sind. Stadtumbau heißt hier „Rückbau", also Abbau von nicht mehr nachgefragtem Wohnungsbestand. Stadtumbau antwortet in dieser Optik auf den dramatischen Einwohnerverlust, auf die „schrumpfende Stadt". Rückbau und städtebauliche Integration der ehemaligen Massenwohnquartiere sind die vielleicht härteste Herausforderung für die europäischen Städte, vor allem, aber nicht nur in Osteuropa.

At the centre of public interest is the redevelopment of the centres of the big cities. There are already signs of where this will lead: the metropolitan centre of the future will be the centre of a suburbanised urban region. Yet not as one fragment among others, but rather as a new kind of centre, one in which suburbanites feel at home. There they will be able to find the unique buildings urgently needed by the less exciting design of the suburban landscape, but also a cleverly thought-out range of high culture, entertainment and rare goods. The new centre is a place of constructed history, on which the faceless suburban landscape is dependent, a destination for tourism, and not only for the continually growing worldwide long-distance tourism, but above all also for suburban tourism on a local scale. In the new centre are congregated select institutions for the production of knowledge, decision-making and creativity. And there is even room and a home for those people who like to live there, including the elderly.

This immediately outlines the urban development programme for redeveloping city centres: the public space is reconquered for pedestrians and made more attractive or redesigned. Spectacular new complexes for entertainment and the arts are built or implanted within historical buildings. There is a better mix of utilisation, above all in the central location of attractive residential space for higher-earners. The centre hums 24-hours a day (the 'round-the-clock metropolis') and the city orients itself – wherever possible – to the water, even of it is only a brackish and not particularly sweetly-smelling canal. The urban greenery is extended and qualified. Finally, the heritage of urban development is carefully preserved and enriched with historicised reconstructions. In addition, there is the attraction of new buildings, which symbolise future perspectives, whether in a deadly serious manner or with more joy.

The second major theme of urban redevelopment concerns 'conversion'. Inside, at the edge of and outside the inner cities, huge areas have become wasteland over the past decades. These include areas formerly used by the railways, industry, trade, the military, trade fairs and airports. How can these be reutilised? Through an extension of the traditional city of through wholly new structures?

Urban redevelopment has recently acquired a third significance, above all in Germany: the adaptation of the large-scale residential estates of industrialised, mass council housing to altered needs. 'Urban Redevelopment East' serves in the first place to demolish those 'slab block buildings' which have lost their residents. Urban redevelopment here means 'rebuilding', i.e. taking down residential buildings that no longer meet a demand. Optically, urban redevelopment is responding to the dramatic loss of residents, to the 'shrinking city'. The rebuilding and urban development integration of former mass residential housing estates are perhaps the most difficult challenge for European cities, above all (but not only) in eastern Europe.

Bernhard Tschumi: Parc de la Villette, Paris

Lyon: Luxusvariante des Umbaus öffentlicher Räume in historischen Zentren während der 1990er Jahre
Lyon: luxury variation on the renewal of public space in historic centres during the 1990s

Stadtumbau - dritte Phase der europäischen Stadterneuerung

Der aktuelle Stadtumbau lässt sich nur verstehen, wenn er in seiner Entwicklung betrachtet wird. Während Stadterneuerung bis Anfang der 1970er Jahre noch als Sanierung im Sinne einer radikalen, autogerechten Neugestaltung, Tertiärisierung und Auflockerung der überkommenen Innenstädte verstanden und praktiziert wurde, geriet diese Konzeption („Kahlschlagsanierung") während der 1970er Jahre in eine Krise und wurde von einer stärker baulich wie sozial erhaltenden Stadterneuerung abgelöst. Für die erste Phase der Kahlschlagsanierung standen Erneuerungsgebiete wie die Gorbals in Glasgow und das Sanierungsgebiet Wedding-Brunnenstraße in West-Berlin, für die zweite Phase der erhaltenden, „behutsamen" bzw. „sanften" Stadterneuerung die historischen Zentren von Bologna und Krakau sowie Stadterweiterungsgebiete aus der Zeit vor dem Ersten Weltkrieg etwa in Wien und Berlin. Das äußerst erfolgreiche Europäische Denkmalschutzjahr 1975 markierte den Höhepunkt dieses radikalen Leitbildwandels.

Um 1980 zeichnete sich eine dritte Phase ab: die auf Neu- bzw. Umbau und Revitalisierung setzende gestaltende Stadterneuerung, die später als Stadtumbau bezeichnet wurde. Stadtumbau betrifft die mehr oder weniger harte gestalterische Anpassung der bestehenden Städte an veränderte Verhältnisse – und zwar in schrumpfenden wie wachsenden Stadtregionen. Der Übergang von der erhaltenden Stadterneuerung zum Stadtumbau ging Hand in Hand mit der Neuformierung der Akteure im Städtebau sowie einer Neudefinition von Verlierern und Gewinnern. Zu den „Verlierern" gehören minderbemittelte Stadtbürger, zu den „Gewinnern" die mittleren Einkommensschichten und private Investoren. Denn die soziale Orientierung des europäischen Städtebaus änderte sich während der 1980er Jahre radikal: Nicht mehr der Widerstand gegen die Verdrängung ärmerer Schichten aus der Innenstadt (ein zentrales Thema der 1970er Jahre) steht im Vordergrund, sondern das Ringen um Verbleib oder Rückkehr besser verdienender Schichten. Die Mittelschichten werden durch die Sirenen des Stadtumbaus umworben – als Einwohner, Konsumenten und Besucher. Dieser Wandel äußert sich auch in neuen Allianzen und Interessen, der Künstlerarchitekt feiert sein Comeback auf Kosten des Partizipationsarchitekten und des Stadtplaners. Dazu kam ein tendenzieller Verlust der Steuerungskraft der öffentlichen Hand zugunsten neuer Formen von public private partnership.

Urban Redevelopment - Third Phase Of European Urban Renewal

The current urban redevelopment process can only be understood if one studies its development. Whereas urban renewal up to the start of the 1970s was still understood and practised as reconstruction in the sense of a radical, car-friendly redesign, tertiarisation and thinning-out of inner cities as they had been before, this concept ('demolition and reconstruction') entered a crisis in the 1970s and was replaced by a type of urban renewal that was more concerned with preservation, both architecturally and socially. For the first phase, the demolition and reconstruction process, areas of renewal such as the Gorbals in Glasgow, or the reconstruction area of Wedding-Brunnenstrasse in West Berlin; for the second phase, the conservational, 'cautious' or 'soft' urban renewal, the historical centres of Bologna and Cracow, as well as urban expansion areas from the period before the First World War, for example those in Vienna and Berlin. The extremely successful European Year of the Preservation of Monuments in 1975 marked the high-point of this radical change of leitmotif.

Around 1980 it became possible to discern a third phase: urban renewal design based on new construction, conversion and renewal, and this was later characterised as urban redevelopment. Urban redevelopment concerns the more or less hard adaptation of the design of existing cities to changed circumstances – namely, in both shrinking and growing urban regions. The transition from conservational urban renewal to urban redevelopment went hand in hand with the reforming of the ranks of those involved in urban development, as well as a redefinition of losers and winners. Among the 'losers' were urban citizens of lower financial means, among the 'winners' the middle-income classes and private investors. After all, the social orientation of European urban development changed radically during the 1980s: it was no longer resistance against the ousting of the poorer classes from the inner city (a central theme of the 1970s) that was in the foreground, but rather the struggle to enable the higher-income classes to either stay or to return. The middle-classes were courted by the sirens of urban redevelopment – as residents, consumers and visitors. This transformation was expressed in new alliances and interests, the artistic architect celebrated his comeback at the cost of the participation architect and urban planner. In addition, there was a directional loss of governmental controlling power in favour of new forms of public and private partnership.

Nicholas Grimshaw:
Druckerei der Financial Times, London Docklands
printing works of the Financial Times, London Docklands

Barcelona: Kultort des Stadtumbaus an der Wende von den 1980er zu den 1990er Jahren
Barcelona: cult location of urban renewal at the transition from the 1980s to the 1990s

„Konversion West": Ingenhoven Overdiek Architekten – Wohnbebauung Duisburg Innenhafen
Urban renewal in in West Germany: Ingenhoven Overdiek Architekten – housing development, Duisburg inner port

Kultereignisse, Kultorte, Kultbücher und Propheten des Stadtumbaus

Wie jede Phase der Stadterneuerung hat auch der Stadtumbau seine großen Themen, Personen und Orte. Als spektakulär inszenierter Auftakt kann die Architekturbiennale in Venedig 1980 angesehen werden, deren Motto – La presenza del passato – Programm war. Große Events wie die Internationale Bauausstellung in West-Berlin, die Olympiade in Barcelona und die Weltausstellung in Lissabon förderten die Neuorientierung des europäischen Städtebaus. Darüber hinaus war und ist das Programm der Kulturhauptstädte Europas ein wichtiges Medium des Stadtumbaus.

War die Architekturbiennale in Venedig die Ouvertüre, so brachte der Stadtumbau von Barcelona den entscheidenden Durchbruch. Wiedergewinnung der vom Autoverkehr beanspruchten öffentlichen Räume, Orientierung der Stadt hin zum Wasser, große Kulturkomplexe zur Revitalisierung von Verfallsgebieten, Mischung der Nutzungen, insbesondere Um- und Neubau von Wohnungen für die Middle Class, und nicht zuletzt ein neuer, traditionalistischer Respekt vor der Geschichte, gepaart mit neomodernistischen Gesten demonstrativer Zukunftsfähigkeit – damit waren die Schlüsselthemen des Stadtumbaus gesetzt, die die Fachdiskussion nicht nur in Europa, sondern auch in Nord- und Südamerika seither beherrschten. Die anderen Kultorte des Stadtumbaus der 1980er Jahre huldigten ebenfalls dieser neuen Programmatik: Paris mit seinen Grands Projets (etwa dem Projekt Grand Louvre) in einer sorgfältig gepflegten traditionellen Stadt, Salzburg, das selbsternannte „Projekt einer europäischen Stadt" mit seinen sorgfältig eingepassten Neubauten und London mit den Großkomplexen um den Bahnhof Liverpool Street. Selbst jenseits des Atlantik wurde umgesteuert: Die Landgewinnung mit dem Aushub des World Trade Center bescherte New York einen neuen Stadtteil, der den Stadtumbau in den USA maßgeblich beeinflussen sollte: Battery Park City in Downtown Manhattan.

Cult Events, Cult Locations, Cult Books And Prophets Of Urban Redevelopment

As in any phase of urban renewal, urban redevelopment, too, has its own great themes, persons and places. As a spectacularly produced start, one might take the Architectural Biennial in Venice in 1980, the motto of which – La presenza del passato – was programmatic. Great events such as the International Architecture Exhibition in West-Berlin, the Olympiad in Barcelona and the world exhibition in Lisbon demanded the reorientation of European urban development. Furthermore, the programme of having cultural capitals in Europe was and still is an important medium for urban redevelopment.

If the Architectural Biennial in Venice was the overture, the urban redevelopment of Barcelona was the decisive breakthrough. The reacquisition of public spaces that had been claimed by car traffic, the orientation of the city towards the water, large-scale arts complexes for the rejuvenation of dilapidated areas, mixed utilisation, in particular the redevelopment and new development of residential housing for the middle class, and not least a new and more traditionalist respect for history, coupled with neo-modernist gestures of demonstrative future ability – these were the key subjects of urban redevelopment and since then they have dominated specialist discussion not only in Europe but also in North and South America. The other cult locations of urban redevelopment in the 1980s likewise paid homage to this new programme: Paris with its grands projets (for example, the Grand Louvre project) in a carefully preserved traditional city; Salzburg with its self-proclaimed 'Project for a European City' and carefully adapted new buildings; and London, with the large-scale complexes around Liverpool Street station. Even beyond the Atlantic new courses were being set: the property obtained from the excavation of the World Trade Center gave New York a new urban area, which the urban development in the USA would significantly influence: Battery Park City in Downtown Manhattan.

London, Canary Wharf

London: Neubauten gegenüber St. Paul's, spektakuläres Beispiel für einen neotraditionalistischen Stadtumbau nach Abbruch modernistische· Bürobauten
London: new buildings opposite St. Paul's, a spectacular example of neo-traditional urban renewal following the demolition of modernist office blocks.

London: Docklands, Canary Wharf, Referenzbeispiel für die Konversion ehemaliger Hafenflächen im Rahmen privaten Städtebaus
London: Docklands, Canary Wharf, reference example for the conversion of former port areas in the framework of private urban construction

In den 1990er Jahren folgten weitere, breit rezipierte Projekte des Zentrumsumbaus. Erinnert sei nur an Lyon mit seinen aufwendigen Platzumgestaltungen und seinem neuen Kulturtempel, an das Wiener MuseumsQuartier, die Revitalisierung von Temple Bar in Dublin, die Fünf Höfe in München, das neue Rathausquartier in Innsbruck, die Umgestaltung der öffentlichen Räume anlässlich des Heiligen Jahres in Rom und nicht zuletzt an den Berliner Stadtumbau, etwa am Beispiel der Neufigurierung des Pariser, Leipziger und Potsdamer Platzes. In London beschleunigte sich der Stadtumbau im Bereich der großen Bahnhöfe, aber auch am Paternoster Square nördlich von St. Paul's Cathedral. Historistische Rekonstruktionen wurden in Moskau durchgeführt und in Berlin geplant.

Zwar hatte Barcelona schon das Thema der neuen Nutzung von Brachflächen aufgeworfen – diese Frage wurde aber in einer anderen Stadt bis ins Detail durchbuchstabiert: In London begann in den 1980er Jahren der Umbau der Docklands, das gewaltigste Konversionsprojekt Europas, das heute schon weit fortgeschritten ist. Das mit der City of London konkurrierende Gebiet der Canary Wharf ist das bekannteste Produkt dieser Megakonversion eines Hafengebietes in einen postindustriellen Stadtteil. Die Docklands konfrontierten die europäische Fachwelt mit einem neuen Phänomen, das inzwischen Alltag geworden ist: mit der zunehmenden Schwächung der öffentlichen Hand im Städtebau. Konversion ist vor allem auch privater Städtebau, aber nicht nur: Der neue Parc de la Villette in Paris war ein Meilenstein öffentlich gesteuerter Umnutzung eines untergenutzten innerstädtischen Areals. Er war zugleich ein Symbol für das zunehmende Gewicht landschaftsplanerischer Maßnahmen im Stadtumbau.

In den 1990er Jahren nahm die Bedeutung von Konversionsprojekten weiter zu. Verwiesen sei auf die Neugestaltung etwa der Hafenflächen in Rotterdam (Kop van Zuid), Cardiff, Genua, Lissabon, Vancouver und Buenos Aires sowie auf den großen Plan einer HafenCity in Hamburg. In der Tübinger Südstadt wurde beispielhaft ein ehemaliges Kasernengelände in ein mischgenutztes Quartier kleinteilig in Regie der öffentlichen Hand transformiert. Eines der größten deutschen Konversionsprojekte hinsichtlich alter Militärflächen findet sich in Potsdam: der Umbau des Bornstedter Felds, der durch eine Bundesgartenschau beschleunigt werden konnte. Ein durchschlagender Erfolg war schließlich die Revitalisierung des untergenutzten Südufers der Themse in London zwischen Westminster und Tower Bridge.

Der Umgang mit den großen Komplexen des Sozialen Wohnungsbaus hat bislang keine vergleichbaren Vorzeigeorte geschaffen. Zwar gab es bereits in den 1980er Jahren zahlreiche Versuche zur Stabilisierung der städtebaulich wie gesellschaftlich prekären Verhältnisse vor allem in französischen, britischen, skandinavischen und deutschen Großsiedlungen. In der deutschen Fachwelt fanden die „Nachbesserung" des Märkischen Viertels in Berlin und Kirchdorf Süd in Hamburg Beachtung. In den 1990er Jahren erregte der abrissorientierte Umbau der Amsterdamer Großsiedlung Bijlmermeer internationales Aufsehen. Die eigentliche Herausforderung stellte sich aber nach dem Fall des eisernen Vorhangs im ehemals sozialistischen Europa: Die Massenquartiere des industrialisierten Wohnungsbaus verloren an Ansehen, ihr Umbau wurde unausweichlich. Der Wettbewerb „Stadtumbau Ost" in den neuen deutschen Bundesländern hat bereits einige Projekte des Rückbaus von „Plattenbausiedlungen" bekannt gemacht, so etwa in Leinefelde, Cottbus, Hoyerswerda oder Schwedt. Und selbst der nachgeschobene „Stadtumbau West"

In the 1990s there followed further, well-received projects of city centre redevelopment. One needs only to remember Lyon and the extensive redesign of its square and its newly-built arts temple, Vienna's MuseumsQuartier, the rejuvenation of Temple Bar in Dublin, the Fünf Höfe in Munich, the new city hall precinct in Innsbruck, the redesign of public space on the occasion of Holy Year in Rome and last but not least the redevelopment of Berlin, for example the reconfiguration of Pariser, Leipziger and Potsdamer Platz. In London urban redevelopment accelerated in the vicinity of the big stations, but also in Paternoster Square north of St. Paul's Cathedral. Historicist reconstructions were implemented in Moscow and planned in Berlin.

Although Barcelona had already opened up the subject of the re-utilisation of wasteland, this question was actually worked out down to the last detail in another city: in London, the conversion of Docklands began in 1984 and was the greatest conversion project in Europe, and it is very advanced even today. The area of Canary Wharf, which now competes with the City of London, is the best-known product of this mega-conversion of a dockland area into a post-industrial urban district. Docklands confronted the European specialist world with a new phenomenon which has, in the meantime, become quite normal: the increasing lack of public funding for urban development. Conversion is, above all, also private urban development, although it is not only that: the new Parc de la Villette in Paris was a milestone of publicly directed reutilisation of an under-used inner city property. It was at the same time a symbol for the increasing weight of landscape planning measures in urban redevelopment.

In the 1990s the significance of conversion projects continued to increased. Reference need only be made the redesign of, for example, the dockland areas in Rotterdam (Kop van Zuid), Cardiff, Genoa, Lisbon, Vancouver and Buenos Aires, or to the large-scale plan for a HafenCity in Hamburg. In an exemplary project in Tübinger Südstadt a former barracks was transformed into an area with mixed utilisation under the direction of public works. One of the biggest German conversion projects involving a military area was in Potsdam, while it was possible to accelerate the redevelopment of Bornstedter Feld through a Federal Gardens show. Finally, the rejuvenation of the unused south bank of the Thames in London between Westminster and Tower Bridge was also a sweeping success.

The approach to the great complexes of council housing has so far not created any comparable showplaces. Although even in the 1980s there were numerous attempts to stabilise the precarious urban development and social conditions on large-scale French, British, Scandinavian and German estates. In the German specialist world the 'adjustments' of the Märkischen Viertel in Berlin and Kirchdorf Süd in Hamburg received a lot of attention. In 1990s the demolition-oriented redevelopment of the large-scale Bijlmermeer estate in Amsterdam caused an international sensation. However, the real challenge came after the fall of the Iron Curtain in former Socialist Europe: the mass estates of industrialised residential housing lost their appeal, making redevelopment inevitable. The competition 'Urban Redevelopment East' in the new Federal German provinces has already publicised several projects for rebuilding the 'slab block estates', for example in Leinefelde, Cottbus, Hoyerswerda or Schwedt. And even the later 'Urban Redevelopment West' is able to present a spectacular example: the rebuilding of the large-scale Osterholz-Tenever estate in Bremen.

Amsterdam: Bijlmermeer, Modellfall des Umbaus einer sozialen Massenwohnsiedlung in den 1990er Jahren
Amsterdam: Bijlmermeer, model case of the renewal of a social mass housing project in the 1990s

Salzburg-Projekt mit Rob-Krier-Entwurf; Thesen von Th. Sieverts
Salzburg-Project with design by Rob-Krier; concepts by Th. Sieverts

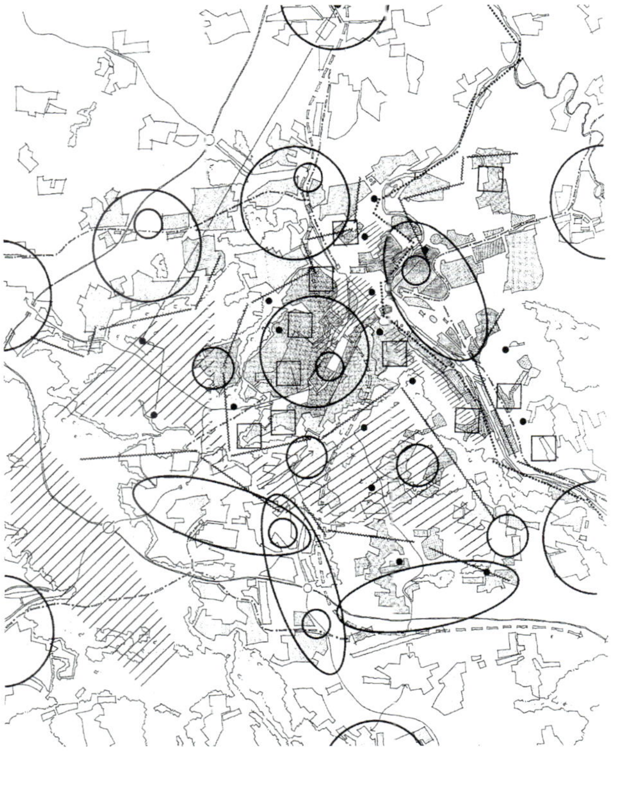

hat ein spektakuläres Beispiel vorzuweisen: den Rückbau der Großsiedlung Osterholz-Tenever in Bremen.
Wie jede Phase des Städtebaus hat auch der Stadtumbau seine geliebten oder weniger geliebten Propheten. Dazu gehören in England Prince Charles und Richard Rogers, in Italien Pier Luigi Cervellati, in Spanien Oriol Bohigas, in Dänemark Jan Gehl, in Holland Rem Koolhaas, in Deutschland Dieter Hoffmann-Axthelm, Andreas Feldtkeller und Hans Stimmann sowie – widerspenstig gegen jede nationale Verortung – der Querdenker Léon Krier. Artikel und Bücher dieser Vordenker werden hochgehalten oder verteufelt, von manchen sogar gelesen, jedenfalls oft zitiert. Beispiele hierfür sind „A Vision of Britain“ (HRH The Prince of Wales, 1989), „Cities for a small planet“ (Richard Rogers, 1997), „Architektur: Freiheit oder Fatalismus“ (Léon Krier, 1998), „Städtebau: Vielfalt und Integration“ (Andreas Feldtkeller, 2001).

Eine neues Thema: Umbau der Stadtregion

„Zwischenstadt“ ist ein kultureller Kampfbegriff, der von dem deutschen Stadtplaner Thomas Sieverts in die Diskussion gebracht worden ist. Zwischenstadt – so nennt Sieverts die Räume zwischen den Städten, Räume, die oft als Suburbia, als Orte der Zersiedelung bezeichnet werden. Es ist das Verdienst von Thomas Sieverts, auf die Vernachlässigung des Themas Suburbia in der fachlichen städtebaulichen Diskussion hingewiesen zu haben. Ist es nicht völlig weltfremd, so Sieverts, der kompakten, europäischen Stadt zu huldigen und zugleich die Entwicklung der suburbanen Peripherie zu verdammen? Ist nicht die Qualifizierung der suburbanen Peripherie die eigentliche Herausforderung für den Städtebau von morgen? Die Debatte, die dieses Buch ausgelöst hat, zeigt deutlich, wie wenig wir eigentlich über Suburbia in Deutschland wissen. Anders als in den USA, England und Holland gibt es bei uns nur sehr vereinzelte Studien über die Entstehung und die Widersprüche des suburbanen Städtebaus.

As in every phase of urban development, urban redevelopment has its popular and less popular prophets. In England they include Prince Charles and Richard Rogers, in Italy Pier Luigi Cervellati, in Spain Oriol Bohigas, in Denmark Jan Gehl, in Holland Rem Koolhaas, in Germany Dieter Hoffmann-Axthelm, Andreas Feldtkeller and Hans Stimmann, as well as – stubbornly against every national localisation – the open-minded thinker Léon Krier. Articles and books by these prophets are either highly regarded or condemned, some people even read them and they are at least often quoted. Examples include: *A Vision of Britain* (HRH The Prince of Wales, 1989), *Cities for a Small Planet* (Richard Rogers, 1997), *Architektur: Freiheit oder Fatalismus* (Léon Krier, 1998), *Städtebau: Vielfalt und Integration* (Andreas Feldtkeller, 2001).

A New Subject: Redevelopment Of The Urban Region

Zwischenstadt ('intermediate city') is a term of cultural warfare, which has been brought into the discussion by the German urban planner Thomas Sieverts. The intermediate city is Sieverts' name for the spaces between the cities, spaces which often are termed suburbia, places of overdevelopment. It is to the credit of Thomas Sieverts that he has pointed out the neglect of the Thames suburbia in the specialist urban discussion. Isn't it completely unworldly, says Sieverts, to pay homage to the compact European city and yet at the same time to condemn the development of the suburban periphery? Isn't the qualification of the suburban periphery the real challenge of the urban development of tomorrow? The debate which this book has triggered clearly shows how little we actually know about suburbia in Germany. In contrast to the situation in USA, England and Holland, only very isolated studies have been made here into the emergence and contradictions of suburban development.

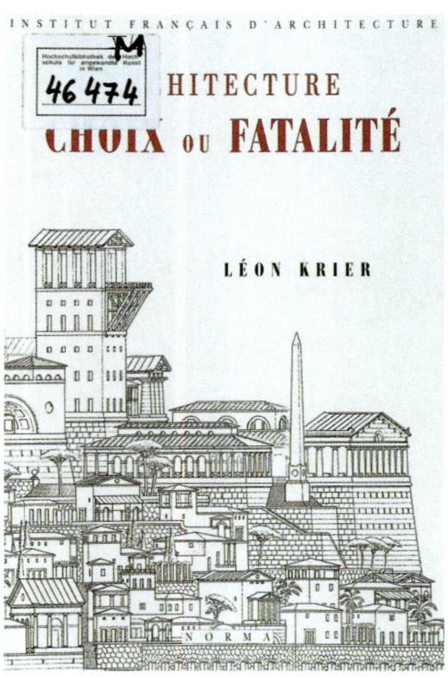

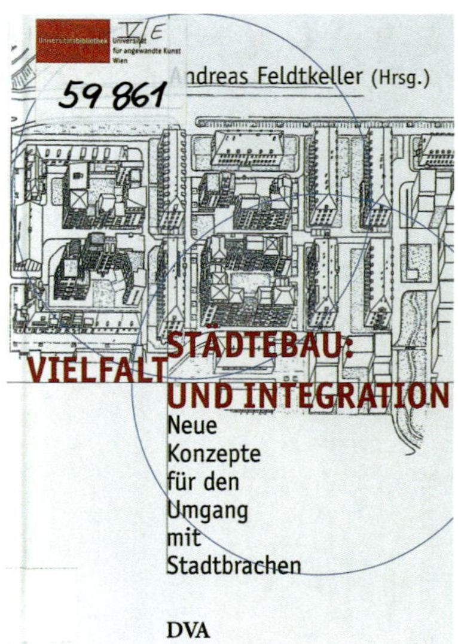

Propheten des Stadtumbaus | prophets of Urban Redevelopment: **Léon Krier und Andreas Feldtkeller**

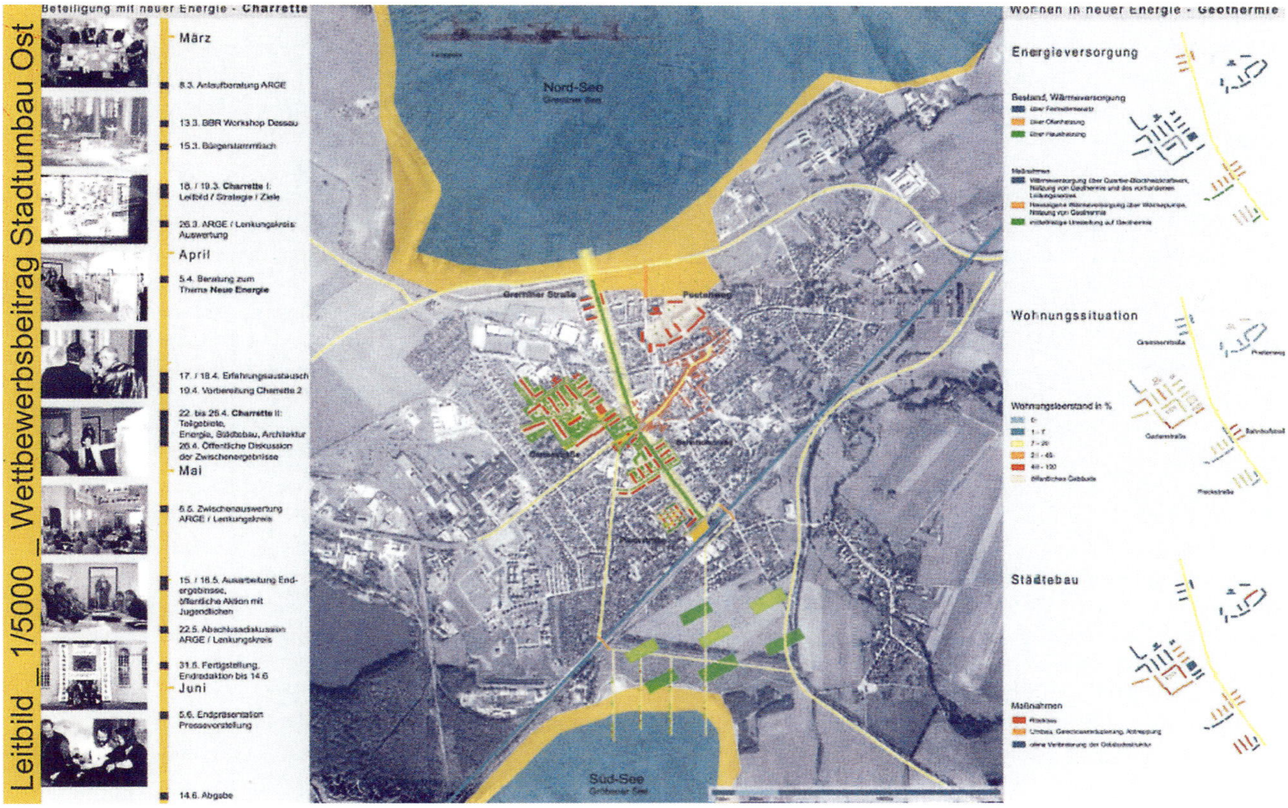

Bundeswettbewerb Stadtumbau Ost: Gräfenhainichen
Urban East German redevelopment competition Gräfenhainichen

Konversion Ost | East German Conversion: **Barkow Leibinger: Biosphäre** | Biosphere**, Potsdam**

Mit den neueren Debatten um Zwischenstadt bzw. Suburbanisierung deutet sich eine vierte Phase der Stadterneuerung an, die dem Stadtumbau eine neue Dimension geben könnte: die Orientierung auf die gesamte Stadtregion und die Überwindung der isolierenden Sicht jeweils nur auf die Innenstadt, auf Konversionsflächen, auf Großsiedlungen oder auf die Gebiete suburbaner Zersiedelung. Ziel wäre perspektivisch der solidarische „Umbau" der gesamten Stadtregion. Der Streit darüber, was das konkret heißt und wo die Prioritäten gesetzt werden, hat allerdings noch gar nicht richtig begonnen. Als erste Beispiele in dieser Richtung können die IBA Emscher Park im Ruhrgebiet und das Langzeitprojekt „Industrielles Gartenreich" in Sachsen-Anhalt gelten.

Wohin diese Debatte programmatisch wie praktisch führen kann, zeigt der Blick über den Atlantik: In den USA hat sich bereits während der 1990er Jahre eine breite städtebauliche Bewegung gegen die Zersiedelung, den „urban sprawl", und für die Revitalisierung der Innenstädte entfaltet. Zumindest programmatisch wurde dort der Gegensatz zwischen den Verteidigern der „kompakten Stadt" und den Propagandisten der „Zwischenstadt" überwunden. Das Buch „The Regional City" von Peter Calthorpe und William Fulton (2001) ist ein Manifest der US-amerikanischen Anti-Sprawl-Bewegung. Ihr Ziel ist nicht die Qualifizierung von Downtown oder Suburbia, sondern die Qualifizierung von Downtown und Suburbia in einem umfassenden Konzept des Umbaus der Stadtregion.

With these latest debates about the intermediate city or suburbanisation, a fourth phase of urban renewal can be discerned, which might be able to endow urban redevelopment with a new dimension: orientation to the whole urban region and transcendence of a more confining view that focuses only on the inner city, conversion areas, large-scale estates or areas of suburban over-development. Viewed in perspective, the aim would be to show solidarity for redevelopment of the whole urban region. However, the quarrel about what that actually means and where the priorities should lie has still not begun properly. IBA Emscher Park in the Ruhrgebiet and the long-term project entitled Industrielle Gartenreich ('Industrial Gardenland') in Sachsen-Anhalt may be regarded as the first examples in this direction.

Where this debate might lead programmatically or practically can be shown by taking a look across the Atlantic: in the USA there unfolded as early as the 1990s a broad urban development movement against over-development or 'urban sprawl' and in favour of the rejuvenation of the inner cities. At least programmatically, the contradiction between the defenders of the 'compact city' and the propagandists of the intermediate city has been resolved. The book The Regional City by Peter Calthorpe and William Fulton (2001) is a manifesto of the US American Anti-Sprawl Movement. Its aim is not the qualification of downtown or suburbia, but rather the qualification of downtown and suburbia as part of a comprehensive concept of redevelopment of the urban region.

Literatur

Becker, Heidede / Jessen, Johann / Sander, Robert (Hrsg.): Ohne Leitbild? Städtebau in Deutschland und Europa. Stuttgart u.a. 1997

Bentley, James: East of the City. The London Docklands Story. London 1997

La Biennale di Venezia, Settore Architettura: La presenza del passato. Prima mostra internazionale di architettura. Venezia 1980

Bohigas, Oriol u.a.: Barcelona. Stuttgart 1991

Calthorpe, Peter / Fulton, William: The Regional City: planning for the end of sprawl. Washington u.a. 2001

Cervellati, Pier Luigi: La città postindustriale. Bologna 1984

La costruzione della città europea negli anni '80. Indagine su come sono cambiate 19 città europee negli ultimi 10 anni. 3 Bände. Roma 1991

Feldtkeller, Andreas (Hg.): Städtebau: Vielfalt und Integration. Neue Konzepte für den Umgang mit Stadtbrachen. Stuttgart / München 2001

Gehl, Jan: Life between Buildings. Using Public Space. O.O. 1996

Hoffmann-Axthelm, Dieter: Die dritte Stadt. Frankfurt am Main 1993

Koolhaas, Rem u.a.: Mutations. 2001

Krier, Leon: Freiheit oder Fatalismus. München 1998

Powell, Kenneth: Stadt im Umbau. Städtebau zu Beginn des 21. Jahrhunderts. Stuttgart München 2000

HRH The Prince of Wales: A Vision of Britain. A Personal View of Architecture. London u.a. 1989. Deutsche Ausgabe: Die Zukunft unserer Städte. Eine ganz persönliche Auseinandersetzung mit der modernen Architektur. München 1990

Rogers, Richard: Cities for a small planet. Edited by Philip Gumuchdjian. Boulder / Colorado 1998

Schilling, Rudolf: Rückbau und Wiedergutmachung. Basel – Boston 1987

Sieverts, Thomas: Zwischenstadt zwischen Ort und Welt, Raum und Zeit, Stadt und Land. 2. Aufl. Braunschweig / Wiesbaden 1998

Steiner, Dietmar (Hrsg.): Das Salzburg-Projekt. Entwurf einer europäischen Stadt. Wien 1986

Stimmann, Hans (Hrsg.): Von der Architektur- zur Stadtdebatte. Berlin 2001

Towards an Urban Renaissance. Final Report of the Urban Task Force. Chaired by Lord Rodgers of Riverside. London 1999

Bibliography

Becker, Heidede/Jessen, Johann/Sander, Robert (Ed.): *Ohne Leitbild? Städtebau In Deutschland Und Europa*. Stuttgart and elsewhere 1997

Bentley, James: *East Of The City. The London Docklands Story*. London 1997

La Biennale di Venezia, *Settore Architettura: La Presenza Del Passato*. Prima Mostra Internazionale Di Architettura. Venice 1980

Bohigas, Oriol And Others: *Barcelona*. Stuttgart 1991

Calthorpe, Peter/Fulton, William: *The Regional City: planning for the end of sprawl*. Washington and other places 2001

Cervellati, Pier Luigi: *La Città Postindustriale*. Bologna 1984

La Costruzione Della Città Europea Negli Anni '80. Indagine Su Come Sono Cambiate 19 Città Europee Negli Ultimi 10 Anni. 3 volumes. Rome 1991

Feldtkeller, Andreas (Ed.): *Städtebau: Vielfalt Und Integration. Neue Konzepte Für Den Umgang Mit Stadtbrachen*. Stuttgart/Munich 2001

Gehl, Jan: *Life between Buildings. Using Public Space*. O.O. 1996

Hoffmann-Axthelm, Dieter: *Die Dritte Stadt*. Frankfurt am Main 1993

Koolhaas, *Rem And Others*: Mutations. 2001

Krier, Leon: *Freiheit Oder Fatalismus*. Munich 1998

Powell, Kenneth: *Stadt Im Umbau. Städtebau Zu Beginn Des 21. Jahrhunderts*. Stuttgart/Munich 2000

HRH The Prince of Wales: *A Vision Of Britain. A Personal View of Architecture*. London and other places 1989.

Rogers, Richard: *Cities For A Small Planet*. Edited by Philip Gumuchdjian. Boulder/Colorado 1998

Schilling, Rudolf: *Rückbau Und Wiedergutmachung*. Basel/Boston 1987

Sieverts, Thomas: *Zwischenstadt Zwischen Ort Und Welt, Raum Und Zeit, Stadt Und Land*. 2nd ed. Braunschweig/Wiesbaden 1998

Steiner, Dietmar (Ed.): *Das Salzburg-Projekt. Entwurf Einer Europäischen Stadt*. Vienna 1986

Stimmann, Hans (Ed.): *Von Der Architektur- Zur Stadtdebatte*. Berlin 2001

Towards An Urban Renaissance. Final Report of the Urban Task Force. Chaired by Lord Rodgers of Riverside. London 1999

South Bank London

South Bank London

Stadtumbau am Südufer der Themse
Urban Renewal On The South Bank Of The Thames

Photos Cordelia Polinna, Christian Richters, Nigel Young
Text Cordelia Polinna

Am Südufer der Themse in London hat sich in den vergangenen fünf Jahren ein Aufwertungsprozess vollzogen, der durch eine Reihe von Neuansiedlungen aus dem Kultur- und Freizeitsektor angestoßen wurde. Noch zu Beginn der 1990er Jahre war das ehemalige Industrie- und Hafengebiet im Norden der Stadtteile Southwark und Lambeth in vielen Teilen isoliert und für Touristen wie Londoner nicht besonders attraktiv. Bis auf historische Sehenswürdigkeiten wie die Tower Bridge oder die Houses of Parliament war London vom Wasser abgewandt, sah den Fluss eher als Transport- und Industrieraum denn als Potenzial für Erholung und Freizeitgestaltung. Doch durch eine Vielzahl von ehrgeizigen Stadtumbauprojekten, darunter der Kunstgalerie „Tate Modern", ist mittlerweile eine Neudefinition des südlichen Themseufers als Touristenattraktion und Kulturmeile gelungen. Vorraussetzungen für diese Entwicklung des Gebietes waren die Verlängerung der U-Bahnlinie Jubilee Line und der Bau zweier Fußgängerbrücken sowie die zentrale Lage gegenüber der „Square Mile", dem Finanz- und Dienstleistungszentrum Londons. Nach jahrzehntelanger Abkehr wendet sich die Stadt wieder dem Fluss zu und entdeckt ihn als identitätsstiftendes Element. Einen „Masterplan" zur Steuerung dieser Entwicklung hat es nicht gegeben; das in zahlreichen Städten erfolgreich beschworene Leitbild der „Stadt am Wasser" ersetzt sehr erfolgreich eine planerische Strategie.

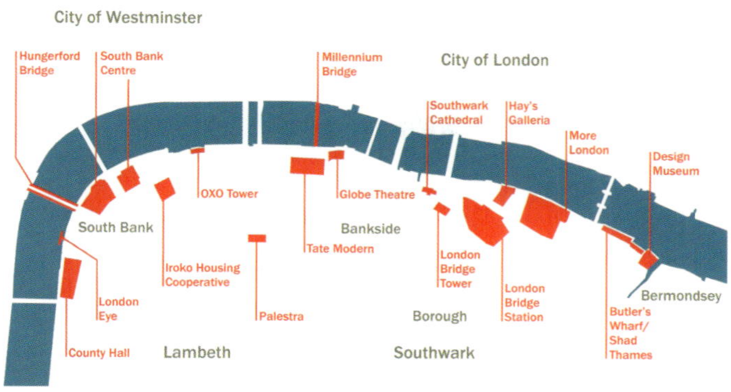

On the south bank of the Thames in London a gradual process of renewal has taken place over the past five years that was set in motion by a series of new developments in the culture and leisure sectors. At the start of the 1990s substantial parts of the former industry and port area in the north of the boroughs of Southwark and Lambeth were isolated and not particularly attractive for either tourists or Londoners. Apart from a few historically important sights, such as Tower Bridge or the Houses of Parliament, London turned away from the water and in general viewed the river as a transport route and industrial zone while ignoring its potential as a recreation and leisure time facility. Thanks to a number of ambitious urban redevelopment projects, among them the Tate Modern gallery, it has proved possible to redefine the south bank of the Thames as a tourist attraction and culture mile. The basic prerequisites for the development of this area were provided by the extension of the Jubilee Underground line, the construction of two pedestrian bridges as well as the district's central location opposite the 'Square Mile', London's finance and service sector centre. After decades of turning away London is now turning towards the river, and discovering it as an element that contributes to shaping the city's identity. There was no 'master plan' to direct this development; the model of the 'city on the water', successfully employed in many other towns, was substituted for a planned strategy.

Das Hinterzimmer Londons

Das Südufer der Themse gilt schon seit mehreren hundert Jahren als ein „Hinterzimmer Londons“. Seit dem Mittelalter blühten in Southwark Theater, Prostitution und preiswerte Gasthäuser, die im religiös reglementierten London nicht geduldet waren. William Shakespeare wirkte in den Theatern Swan, Globe und Rose, die wegen Unsittlichkeit jedoch 1642 geschlossen und abgerissen wurden.

Im 18. Jahrhundert legten Glasherstellung, Schiffbau und Werften den Grundstein für eine industrielle Nutzung des Gebietes, die sich im 19. und beginnenden 20. Jahrhundert noch verstärken sollte. Die Bahnhöfe Waterloo und London Bridge wurden errichtet und wuchsen schnell zu überregional bedeutsamen Verkehrsknotenpunkten. Mittlerweile hatte sich das Südufer der Themse zu einem wichtigen Standort der Lederverarbeitung und Lebensmittelindustrie entwickelt, an den Arbeiterwohngebiete, in der für Großbritannien typischen Reihenhausstruktur, angrenzten.

Der Niedergang von Hafen und Industrie nach dem Zweiten Weltkrieg traf die Boroughs Southwark und Lambeth hart. Im 1943 verabschiedeten Abercrombie-Plan für den Wiederaufbau Londons war vorgesehen, die innerstädtischen Industriestandorte nicht in Stand zu setzen. Der Plan verfolgte das Leitbild der Dezentralisierung. Arbeitsplätze und Wohnstandorte wurden an den Rand der Stadt umgesiedelt. In Hinblick auf die Anforderungen der modernen Schifffahrtstechnologie und Logistik wurden Hafennutzungen flussabwärts, z.B. nach Tilbury, verlagert. Die Lagerhäuser und Zulieferbetriebe in den innenstadtnahen Gewerbegebieten verloren ihre Funktion; benachbarte Quartiere verödeten auf Grund der Verschlechterung der wirtschaftlichen Situation der Bevölkerung. Somit stand die Stadtverwaltung vor der schwierigen Aufgabe, neue Perspektiven zu entwickeln, was zunächst nur eingeschränkt gelang.

Erste Revitalisierungsversuche

Das „Festival of Britain“ (1951) war ein erster Versuch, den besonders stark zerstörten Bereich zwischen County Hall und Waterloo Bridge zu revitalisieren. Eine große Ausstellung versuchte, den wirtschaftlichen Aufbruch der Nation nach dem Krieg zu veranschaulichen, die erhoffte internationale Ausstrahlung blieb jedoch aus. Neben der Royal Festival Hall entstand ein „Dome of Discovery“, der nach der Ausstellung abgerissen wurde. Auf dem Areal wurden später die Queen Elizabeth Hall (1964), die Hayward Gallery (1964) und das Royal National Theatre (1967-77) errichtet. In den folgenden Jahren entwickelte sich das unter dem Namen South Bank Centre bekannte Ensemble zu Großbritanniens renommiertestem Kulturzentrum.

London's Back Room

For several centuries the south bank of the Thames has been regarded as London's back parlour. Since mediaeval times theatre, prostitution and reasonably priced inns flourished in Southwark, facilities that were not generally tolerated in the more puritan atmosphere of London proper. William Shakespeare was active in the Swan, Globe and Rose theatres, which were finally closed down in 1642 and demolished.

In the 18th century glass blowing, shipbuilding and dockyards formed the basis for the industrialisation of the area, a process that was to accelerate in the 19th and early 20th centuries. Waterloo and London Bridge railway stations were built and rapidly grew into important centres in the interregional transport network. In the meantime the south bank of the Thames had become an important centre for leather processing and the foodstuffs industry, bordered by areas of typical London working class terrace housing.

The decline of the port and of industry after the Second World War hit the boroughs of Southwark and Lambeth particularly severely. In the Abercrombie plan for the reconstruction of London, passed in 1943, it was decided not to renew inner urban industrial locations. This plan pursued the concept of decentralisation: workplaces and housing were moved to the periphery of the city. In view of the demands of modern shipping technology and logistics the port facilities were moved further downstream, e.g. to Tilbury. Warehouses and supply industries in inner city industrial areas lost their function, neighbouring areas decayed due to the decline in the economic status of the local population. So the municipal authorities were confronted with the difficult task of developing new perspectives; their initial success in doing this was extremely limited.

First Attempts At Revitalisation

The Festival of Britain (1951) was a first attempt at revitalising the area between County Hall and Waterloo Bridge, which had been particularly badly damaged. This major exhibition attempted to demonstrate the economic revival of the nation after the war, but the international resonance hoped for was not achieved. In addition to the Royal Festival Hall a Dome of Discovery was built which was demolished after the exhibition. On this site the Queen Elizabeth Hall (1964), the Hayward Gallery (1964) and the Royal National Theatre (1967-77) were later erected. In the years that followed this ensemble, known as the South Bank, developed into Britain's best-known culture centre.

Park Street

Bürogebäude aus den 1960er Jahren
1960s office building

Norman Foster: Millennium Bridge

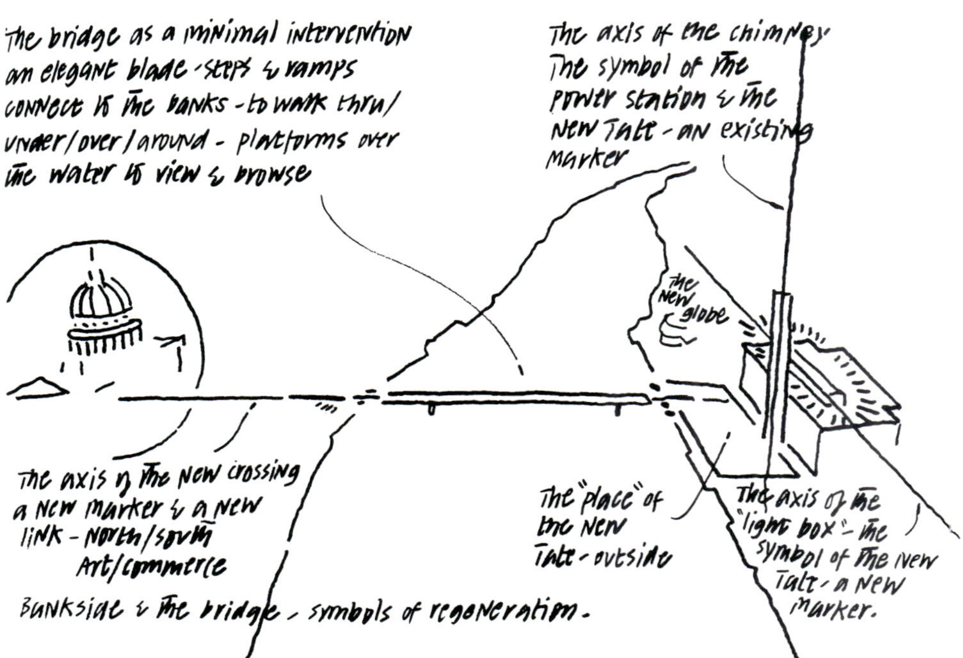

Als Vordenker für eine Entwicklung der Themseufer als Kultur- und Erholungsraum kann Richard Rogers bezeichnet werden, der seit Mitte der 1980er Jahre kontinuierlich für eine Revitalisierung plädierte. Doch bis zum Beginn der 1990er Jahre versuchte das Borough of Southwark, spekulative Bauprojekte und Einrichtungen, die „Yuppies" anziehen würden, von dem Gebiet fernzuhalten. Der Bau von Wohnungen und sozialer Infrastruktur für die ortsansässige Arbeiterbevölkerung wurde gefördert, um Verdrängungsprozesse zu vermeiden. Die zunehmende Nachfrage nach konvertierten Lager- und Hafengebäuden von Nutzern aus der Kultur- und Medienbranche während des Immobilienbooms in den 1980er Jahren und der mit den Kommunalwahlen im Jahre 1992 verbundene Generationswechsel in der Verwaltung führten zu einem Wandel der planerischen Zielsetzung. In den folgenden Jahren bemühte sich der Bezirk sogar selbst um die Ansiedlung von Kultur- und Freizeiteinrichtungen.

Tourismus-Magneten

Das Südufer der Themse ist heute in höchstem Maße heterogen. Das im Zentrum der Aufmerksamkeit stehende Areal erstreckt sich über die Stadtteile South Bank, Bankside, Borough und Bermondsey. Der Stadtteil Borough um die Southwark Cathedral, Clink und Stoney Street ist wegen der noch gut erhaltenen historischen Bausubstanz inzwischen beliebte Tourismusdestination und erlangte als Kulisse zahlreicher Filmproduktionen, z.B. Harry Potter, einen hohen Bekanntheitsgrad. In Bankside und im Umfeld der Station Waterloo dominieren gestalterisch wenig anspruchsvolle Bürogebäude aus der Nachkriegszeit, die als Backoffices für Firmen in der City fungieren. Das gesamte Gebiet wird von Eisenbahnviadukten durchschnitten, entlang derer sich eine Zone aus produzierendem Gewerbe erstreckt. Stark befahrene Straßen, unattraktive Erdgeschoßnutzungen und undurchdringbare Büroblöcke verstärken die Fragmentierung der räumlichen Struktur. Im Kontrast dazu stehen einige kleinteilige Reihenhaussiedlungen z.B. an der Cornwall Street, die in dieser zentralen Lage völlig deplaziert wirken.
Aus Protest gegen die großflächigen Büroprojekte, speziell gegen ein von Richard Rogers entworfenes Bürogebäude, gründete sich 1984 die Genossenschaft Coin Street Community Builders. Seitdem verwirklichte sie eine Reihe von Vorhaben, z.B. den Bau von 220 Wohnungen in Genossenschaftseigentum oder die Konversion des OXO Tower, zuvor eine Brühwürfelfabrik, in Galerien, Wohnungen und Restaurants. Die Projekte entwickelten sich schnell zu Tourismusmagneten. Dies nützt auch der lokalen Bevölkerung, da das gemeinnützige Unternehmen seine Gewinne dazu nutzt, nicht gewinnbringende Aktivitäten zu subventionieren. Die

Richard Rogers can be described as one of the masterminds of the development of the Thames banks as an area for culture and recreation facilities. Since the mid-1980s he has continually argued for its revitalisation. Yet until the beginning of the 1990s the Borough of Southwark attempted to keep speculative development projects and facilities that would attract 'yuppies' to the area at bay. The erection of apartments and a social infrastructure for the local working class population was encouraged in order to combat forces driving such people away from the area. The increasing demand for converted warehouse and port buildings on the part of users from the culture and media sectors during the property boom of the 1980s, combined with a generation change in the political administration linked to the local elections in 1992, led to a change in planning goals. In the years that followed even the district itself attempted to attract culture and leisure time facilities.

Tourist Magnets

The south bank of the Thames is today extremely heterogeneous; the area focussed on here extends across the districts of South Bank, Bankside, Borough and Bermondsey. The district of Borough around Southwark Cathedral, Clink and Stoney Street has now become a popular tourist destination due to its well-preserved historic buildings and has even achieved a certain degree of fame as a backdrop for numerous film productions, e.g. the Harry Potter series. Bankside and the area around Waterloo Station are dominated by unprepossessing post-war office buildings that function as back offices for companies based in the City. The entire district is sliced by a railway viaduct lined by a zone of manufacturing industry. Busy streets, unattractive ground floor uses and impenetrable office blocks increase the fragmentation of the spatial structure. In contrast, there are a number of small-scale terrace housing developments such as that on Cornwall Street, which seem completely out of place in this central location.
The Coin Street Community Builders association was founded in 1984 as a protest against the large office projects, in particular against an office building designed by Richard Rogers. Since its foundation it has realised a series of projects, for example the construction of 220 apartments owned by cooperatives or the conversion of the OXO Tower, formerly the manufacturing place of the stock cube, into galleries, apartments and restaurants. The projects soon developed into tourist attractions. This also benefits the local population as the communally oriented association uses its profits to subsidise non-profit making activities. The architecture of the Coin Street Community Builders differs

South Bank Centre, Bestand
South Bank Centre, existing structure

Thames Walkway

South Bank Centre
Masterplan: Rick Mather Architects

Photos Virtual Artworks

Architektur der Coin Street Community Builders hebt sich deutlich von den britischen Standards im mittelpreisigen Wohnungsbau ab. Es erstaunt jedoch der Maßstab der Anlage, der für eine Kleinstadt angemessen sein mag, nicht aber für eine Millionenstadt wie London, die permanent über ihr Verkehrs- und Pendlerproblem sowie chronische Wohnungsnot klagt. Trotz der innerstädtischen Lage ist beispielsweise der 2001 errichtete Wohnblock „Iroko" von Haworth Tompkins Architects aus London nur drei- und viergeschoßig, die Wohnungen haben sogar kleine Privatgärten.

Mit dem Ende der 1990er Jahre fertiggestellten Thames Path wurde ein Erlebnisweg geschaffen, auf dem man am Südufer entlang der neuen Sehenswürdigkeiten flanieren kann. Der Besuch der Themse wird zu einem Gesamterlebnis, das sich kein Tourist mehr entgehen lassen darf. Der Weg, der nicht nur an interessanten Einrichtungen, sondern auch an künstlerisch und landschaftsarchitektonisch gestalteten Abschnitten sowie einigen „Piazze" entlangführt, kommt dem Wunsch nach einem „mediterranen Lebensstil" der Bewohner und Planer entgegen. Auch wenn die Sommersaison in London nur kurz und es zum „Flanieren" meist zu regnerisch ist, will die Stadt, deren Image noch vor wenigen Jahren untrennbar mit Smog und Nebel verknüpft war, nun südeuropäisches Flair entwickeln. Schon ab Temperaturen von 12° C werden die „open air"-Angebote genutzt.

Die im Februar 2002 wiedereröffnete Millennium Bridge von Foster and Partners verbindet den Thames Path mit der City und lässt die Entfernungen vor allem zu St. Paul's Cathedral schrumpfen. Auch von etablierten Tourismus Highlights, wie dem West End, ist das Südufer der Themse schnell über die neue Hungerford Footbridge zu erreichen. Noch skeptische Besucher werden mit dem Riverside Bus vom Tourismusmagneten Covent Garden abgeholt. Diese prestigeträchtige Buslinie fährt entlang des Südufers die wichtigsten Touristenattraktionen an und verbindet viele Destinationen miteinander.

Das Angebot von Einrichtungen der Hoch- und Populärkultur wird von in- und ausländischen Besuchern sowie von Angestellten während der Mittagspause wahrgenommen. Teilbereiche des Ufers, die noch vor wenigen Jahren zu den am stärksten verwahrlosten Gegenden Londons zählten, z.B. die Clink Street, sind jetzt Fixpunkte auf dem touristischen Stadtplan. Dort gibt es sogar schon ein Starbucks Café, ein untrügliches Zeichen dafür, dass ein Ort in der globalisierten Welt des Städtetourismus angekommen ist.

greatly to the prevailing standard in British medium-price housing. The scale of the development is unusual and seems more appropriate to a small town than to a city such as London with a population of millions, which is permanently faced with traffic and commuter problems as well as a chronic housing shortage. Despite the inner city location the Iroko housing block, built in 2001 by Haworth Tompkins Architects from London, for example, is only three to four storeys high and the apartments even have small private gardens.

The Thames Path, completed at the end of the 1990s, created a route along which one can stroll past the new sights on the south bank. A visit to the Thames has become a complete experience that no tourist wants to miss. This route not only leads past interesting facilities but also boasts stretches that reveal an input by artists and landscape planners along with several 'piazzas'. In general it satisfies a wish for a Mediterranean lifestyle expressed by residents and planners. Although the summer in London is short and it is generally too rainy to stroll, the city, whose image was inseparably linked to smog and fog only a few years ago, wishes to develop a southern European flair. People begin to use the open-air facilities at temperatures of only 12° C!

The Millennium Bridge by Foster and Partners, reopened in February 2002, links the Thames Path to the City and reduces the distance to St. Paul's Cathedral. The south bank of the Thames can now also be reached quickly from established tourist highlights such as the West End via the new Hungerford Footbridge. Visitors who remain sceptical can be collected by the Riverside Bus from the tourist attraction Covent Garden. This prestigious bus line travels along the south bank to the most important tourist sights and connects many different tourist destinations.

The wide range of facilities for classical and popular culture is enjoyed by visitors from Britain and abroad, as well as by office workers during their lunch break. Parts of the riverbank, such as Clink Street, which only a few years ago were among the most rundown areas in London, are now fixed destinations on the tourist map. There is even a Starbucks Café, an unmistakeable sign that a place has finally 'arrived' in the globalised world of urban tourism.

The London Eye

Hungerford Footbridge

OXO Tower; „Broadwall" Estate (Lifschutz Davidson Architects)

Norman Foster:
Greater London Authority Building, London City Hall

Konversionsprojekte – Pioniere des Stadtumbaus
Als wahre Publikumsmagnete haben sich die Projekte erwiesen, die historische Gebäude umnutzen und mit dem maritimen Flair der Anlagen spielen, z.B. die Hay's Galleria (1986) und der OXO Tower (1996). Als Pionier des Stadtumbaus erwies sich das Design Museum von Terence Conran in Butler's Wharf, das seit seiner Eröffnung 1989 als Motor für die Revitalisierung des gesamten Areals östlich der Tower Bridge fungierte. Im Zentrum der Aufmerksamkeit steht die aufwändig umgestaltete Straße Shad Thames, die durch ihre canyonartige Struktur und den Docklands-Charakter besticht. Die konvertierten Lagerhäuser wurden um einige markante Neubauten ergänzt, z.B. den Apartmentblock „The Circle" von CZWG Architects (1989). In schnellem Tempo werden weiter östlich liegende Warehouses in Luxusappartments umgewandelt. Der „Big Bang" für die Revitalisierung kam jedoch im Jahr 1994 mit der Ankündigung des Tate-Direktors Nicholas Serota, eine Galerie für zeitgenössische Kunst in der leerstehenden Bankside Power Station unterbringen zu wollen. Gefördert wurde diese Entscheidung durch die Bereitstellung von £ 50 Mio. aus dem Millennium Fund der staatlichen Lotterie. Der von 1955 bis 1963 von dem Architekten Sir Gilbert Scott errichtete imposante Stahlskelettbau mit Backsteinverkleidung bietet optimale Bedingungen für die Präsentation großformatiger Objekte und publikumswirksamer „blockbuster"-Ausstellungen, die in London bis dahin nicht vorhanden waren.

Conversion Projects – Pioneers Of Urban Renewal
It is the conversions of historic buildings that utilise the maritime flair of the location which have turned out to be true magnets for the general public. Examples include Hay's Galleria (1986) and the OXO Tower (1996). The Design Museum by Terence Conran in Butler's Wharf, which since it opened in 1989 has functioned as a motor for the revitalisation of the entire area to the east of Tower Bridge, is a pioneer in the area of urban renewal. The expensively redesigned Shad Thames Street is a focus of attention; its canyon-like structure and docklands character are particularly impressive. The converted warehouses were complemented by a number of new buildings such as The Circle apartment block by CZWG architects (1989). Further east, warehouses are being converted into luxury apartments at a rapid pace. However the 'Big Bang' for the revitalisation came in 1994 with the announcement by Tate Director Nicholas Serota, of his intention to locate a gallery for contemporary art in the disused Bankside Power Station. This decision was encouraged by the provision of 50 million pounds sterling from the Millennium funds of the State Lottery. The imposing brick-clad steel-frame building, erected from 1955 to 1963 by Sir Gilbert Scott, offers optimal conditions in which to show large-scale pieces and hold popular 'block-buster' exhibitions, which did not previously exist in London.

Norman Foster: London City Hall

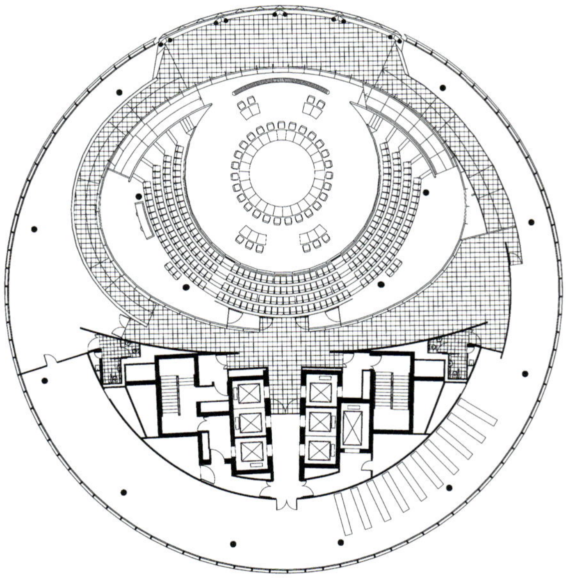

In May 2000 the Tate Modern was opened, remodelled under the leadership of the Swiss architects Herzog & de Meuron, and attracted over three million visitors in its first year. Since then the Tate Foundation has been able to play in the international league of major art museums alongside the Guggenheim Foundation, and is certain of attracting the attention of a growing number of art interested tourists. The extensive reporting of this new opening helped Bankside become known worldwide and attracts tourists to previously almost unexplored regions.
In addition to the conversion projects there has also been a series of major projects with a strong public profile in the areas of culture and leisure, such as the reconstruction of Globe Theatre, as well as the Millennium Project.

Millennium Projects

The Millennium projects, which were financed by foundations and public funds and which include the Tate Modern, serve as a catalyst for the further development of the area and is creating an environment that is attractive to private investment. Despite initial teething trouble they have greatly increased the success of the development of the south bank. The Millennium celebrations were thought of as a symbol of a new beginning following the Thatcher era and the recession of the early 1990s, and were seen as a departure from Great Britain's traditional and somewhat

Im Mai des Jahres 2000 wurde die unter der Leitung des schweizerischen Architekturbüros Herzog & de Meuron umgebaute Tate Modern eröffnet und zog schon im ersten Jahr über drei Millionen Besucher an. Die Tate Foundation kann seitdem in der internationalen Liga großer Kunstmuseen neben der Guggenheim Foundation mitspielen und ist sich der Aufmerksamkeit einer wachsenden Zahl von kulturinteressierten Touristen gewiß. Die umfangreiche Berichterstattung über die Neueröffnung verhalf Bankside zu weltweiter Bekanntheit, und lockt Touristen in bis dahin nahezu unerschlossene Gebiete.
Zusätzlich zu den Konversionsprojekten entstanden eine Reihe von publikumswirksamen Großprojekten aus dem Kultur- und Freizeitsektor, z.B. die Rekonstruktion des Globe Theatre, sowie die Millennium Projects.

Millennium Projects

Die durch Stiftungen oder mit öffentlichen Geldern finanzierten Millennium Projects, zu denen auch die Tate Modern zählt, wirken als Katalysator für die weitere Entwicklung des Gebietes und schaffen ein Umfeld, das private Investitionen anzieht. Trotz einiger Startschwierigkeiten potenzierten sie den Erfolg des südlichen Themseufers. Die Millenniumsfeierlichkeiten waren als Symbol für Neubeginn und Aufbruch nach der Ära Thatcher und der Rezession der frühen 1990er Jahre gedacht und verstanden sich als Abschied vom traditionellen, etwas verstaubten Image Großbritanniens. Die Vorhaben sollten das von Blair ausgerufene „New Britain" baulich umsetzen. London sollte im Mittelpunkt der Aufmerksamkeit stehen und als „coole" Stadt die jungen Kreativen aus aller Welt begeistern.
An der Themse wurden als Millennium Projekte die Tate Modern, die Hungerford Footbridge zwischen Charing Cross und South Bank Centre (Lifschutz Davidson), ein neues Besucherzentrum der Southwark Cathedral, die Millennium Bridge (Foster and Partners) zwischen St. Paul's Cathedral und der Tate Modern und das Riesenrad BA London Eye (Marks Barfield Architects) realisiert. Weiter östlich gelegen befindet sich das ehrgeizigste Projekt, der gigantische Millennium Dome (Richard Rogers Partnership).
Die Vorhaben konnten den hohen Anforderungen nur zum Teil gerecht werden. Fast tragisch anmutende Ereignisse stellten das Visionäre der Projekte und das Streben nach technischer Perfektion in Frage: Das Riesenrad konnte zunächst nicht aufgerichtet werden, und die Millennium Bridge wackelte so stark, dass sie für mehrere Millionen Pfund verstärkt werden musste. Der Millennium Dome wurde so lange von der Presse schlecht geredet, dass letztendlich kaum noch jemand die aufwändig gestaltete Ausstellung besichtigen wollte. Mit Ausnahme des Dome gehören die Anlagen inzwischen zu den beliebtesten Touristenattraktionen.
Das südliche Themseufer wird sich in den kommenden Jahren noch stärker zu einem kulturellen Zentrum wandeln, denn im Frühjahr 2003 eröffnete der Kunstmäzen und Förderer der „Young British Artists" Saatchi in der ehemaligen County Hall an der Westminster Bridge eine Galerie für zeitgenössische britische Kunst. Auch die weiter westlich gelegene Battersea Power Station wird ab Ende 2003 durch Geoffrey Reid Architects, Nicholas Grimshaw und Arup Associates u.a. in eine „gallery of creative business" umgebaut.

Das South Bank Center

Die Zunahme an hochkarätigen Kultureinrichtungen und der damit einhergehende Anstieg der Besucherzahlen übt Druck auf schon bestehende Institutionen aus. So fürchtet z.B. das South Bank Centre um seine internationale Konkurrenzfähig-

fusty image. The plans were intended to offer a built image of the 'New Britain', called into life by Tony Blair. London was to be at the centre of attention and, as a 'cool' city, become a focus for the enthusiasm of creative young people from all around the world.
The Millennium projects on the Thames include the Hungerford Footbridge between Charing Cross and the South Bank Centre (Lifschutz Davidson), a new visitors centre for Southwark Cathedral, the Millennium Bridge between St. Paul's Cathedral (Foster and Partners), the Tate Modern and the British Airways London Eye observation wheel, (Marks Barfield Architects). The most ambitious project, the gigantic Millennium Dome (Richard Rogers Partnership), lies further east.
These projects were not entirely able to meet the great expectations made of them. Almost tragic circumstances called into question the visionary nature of the projects and their drive for technical perfection: the observation wheel could initially not be erected and the Millennium Bridge was so shaky that it had to be reinforced at a cost of several million pounds. The Millennium Dome was criticised in the press for so long that ultimately hardly any one wanted to visit the expensively designed exhibition there. However, with the exception of the Dome, all these facilities are now among the most popular tourist attractions.
In the years to come the south bank of the Thames will increasingly transform into a cultural centre as in spring 2003 Saatchi, the art patron and supporter of Young British Artists, opened a gallery for contemporary British art in the former County Hall at Westminster Bridge. Also, further to the west, Battersea Power Station is to be converted from the end of 2003 by Geoffrey Reid Architects, Nicholas Grimshaw and Arup Associates (among others) into a "gallery of creative business".

The South Bank Centre

The growing number of high-quality culture facilities and the associated increase in the number of visitors are exerting a certain pressure on existing facilities. The South Bank Centre, for example, is concerned about maintaining its international position as hardly any renewal work has been carried out since the completion of this complex at the end of the 1970s. Previous attempts failed at the initial planning phase. Today the lack of openness onto the river, the poor accessibility and the condition of the outdoor facilities mean that the complex no longer meets current requirements. The fittings of the buildings themselves must be adapted to meet modern technical standards.
London architect Rick Mather emerged as the winner of a competition held in 1999 for the redesign of the ensemble. The central points of his master plan are the renovation of the Royal Festival Hall, the extension of the Hayward Gallery as well as the redesign of Jubilee Gardens and the adjacent car park. An elevated park is to be created that rises to a height of 20 metres from the Thames and beneath which a further concert hall with a capacity of 1,500 seats will be built. Existing blank concrete facades will be replaced by glass fronts. The undefined outdoor spaces dominated by delivery traffic are to be replaced by so-called 'active building frontages', i.e. accessible ground floor zones which should make this a much more pleasant place to linger.
At present the master plan is being more precisely defined by further architecture competitions. The redesign of the Royal Festival Hall is being carried out by the London practice of Allies and Morrison. The architects are adding a café to the previously ignored rear (south) side

Herzog & de Meuron: Tate Modern

Butler's Wharf/Shad Thames

Great Suffolk Street:
Eisenbahnviadukt auf dem Weg zu Tate Modern
Railway viaduct on the way to the Tate Modern

keit, denn seit der Fertigstellung des Komplexes Ende der 1970er Jahre wurden kaum noch Erneuerungsmaßnahmen durchgeführt. Versuche scheiterten bisher immer schon zu Beginn der Planungsphase. Inzwischen entsprechen die mangelnde Öffnung der Gebäude zum Fluß, die Zugangsmöglichkeiten und der Zustand der Außenanlagen nicht mehr den heutigen Ansprüchen. Auch die Ausstattung der Gebäude muss neuen technischen Standards angepasst werden.
Aus einem 1999 ausgelobten Wettbewerb zur Umgestaltung des Ensembles ging der Londoner Architekt Rick Mather als Sieger hervor. Kernpunkte seines Masterplans sind die Renovierung der Royal Festival Hall, die Erweiterung der Hayward Gallery sowie die Umgestaltung der Jubilee Gardens und des angrenzenden Parkplatzes. Dort soll ein Park entstehen, der von der Themse aus bis auf eine Höhe von 20 m ansteigt und unter dem ein weiterer Konzertsaal mit 1.500 Plätzen liegen wird. Die blanken Betonfassaden werden durch Glasfronten ersetzt. Statt der undefinierten und vom Lieferverkehr dominierten Außenräume wird es in Zukunft sogenannte active building frontages, also zugängliche Erdgeschoßzonen, geben, die die Aufenthaltsqualität stark verbessern sollen.
Derzeit wird der Masterplan durch weitere Architekturwettbewerbe konkretisiert. Die Umgestaltung der Royal Festival Hall wird vom Londoner Büro Allies and Morrison durchgeführt. Die Architekten fügen auf der südlichen bisher vernachlässigten Rückseite des Gebäudes ein Cafè ein und definieren so den angrenzenden Platze neu.

More London

Nachdem der Imagewechsel des Flussufers zunächst durch Kulturprojekte, vor allem als Konversion historischer Industriegebäude, vorbereitet wurde, werden dort in naher Zukunft eine Reihe von großflächigen Büroprojekten entstehen. Allein im Umfeld der London Bridge Station, die als Verkehrsknotenpunkt für die Ansiedlung von Dienstleistungsunternehmen prädestiniert ist, stehen drei Großprojekte am Beginn bzw. Ende der Realisierung. Die Umgestaltung der London Bridge Station, die Errichtung des von Renzo Piano geplanten London Bridge Tower und das schon fast fertiggestellte More London Projekt an der Tooley Street werden dem Gebiet 37.000 weitere Arbeitsplätze und somit auch Kunden für entsprechende Gastronomie- und Shoppingangebote bescheren.
Das kommerzielle Büroquartier „More London“ nach Plänen des Büros Foster and Partners westlich der Tower Bridge ist von besonderer symbolischer Bedeutung für die Aufnahme Southwarks in den Kreis der Vorzeigegebiete Londons, denn es beinhaltet das neue Rathaus. Nach 14-jähriger Abstinenz wurden im Jahr 2000 infolge von Beschlüssen der Labour-Regierung wieder ein Stadtparlament (Greater London Authority, GLA) und ein Bürgermeister (Ken Livingstone) gewählt.
Auf dem insgesamt 5,5 Hektar großen Gelände werden bis 2005 270.000 m^2 Geschoßfläche zur Nutzung durch Büros, ein Hotel, Einzelhandel, Gastronomie und das Rathaus errichtet. Die Gebäudestruktur entwarfen Foster and Partners zusammen mit der Firma Space Syntax, sie leitet sich aus dem Verlauf der idealen Fußgängerverbindung zwischen der London Bridge Station und der Tower Bridge ab. Wichtige Komponente des städtebaulichen Entwurfs ist eine Reihe von Piazze, ohne die kaum noch eines der zahlreichen neuen Projekte im Zentrum der Stadt auskommt. Die Aufenthaltsqualität der öffentlichen Räume wird durch „lunchtime entertainment“ sowie Ruhezonen speziell den Bedürfnissen der Hauptzielgruppe der Büroangestellten angepasst. Für eine

of the building and so redefining the adjoining outdoor space.

More London

Following the transformation of the riverbank initiated by cultural projects, above all in the form of conversions of historic industrial buildings, in the near future a series of large-scale office projects is to be erected there. Alone in the vicinity of London Bridge Station – a transport node point predestined as a location for service industries – three major projects are about to start or are nearing completion. The redesign of London Bridge Station, the erection of London Bridge Tower planned by Renzo Piano and the almost completed More London project on Tooley Street will provide the area with 37,000 further workplaces, providing customers with appropriate restaurant and shopping facilities.
The More London commercial office district to the west of Tower Bridge, planned by Foster and Partners, has a particularly symbolic importance for the acceptance of Southwark in the circle of London's 'show districts' as it contains the city's new Town Hall. In 2000, following a 14-year abstinence, the Labour Government decided to hold elections for a municipal council (Greater London Authority, GLA) and the position of mayor (Ken Livingstone).
By the year 2005 on a site totalling 5.5 hectares in area, 270,000 square metres of floor area comprising offices, hotels, retail units, restaurants and the Town Hall is to be created. The building structure was designed by Foster and Partners in collaboration with the Space Syntax company. It is derived from the course of the ideal pedestrian connection between London Bridge Station and Tower Bridge. An important component of the urban design is a series of piazzas, an almost ubiquitous feature in the numerous new city-centre projects. The quality of these public spaces as a place to linger is adapted to meet the requirements of the main target group, the office workers, by the provision of 'lunchtime entertainment' as well as quiet areas. The central underground vehicular access system makes the space pedestrian friendly.
One particular example of public-private partnership is the renting of the privately built Town Hall by the Greater London Authority. This egg-shaped, completely glazed building contains an assembly room, committee rooms, offices and a café. Important features include the circular plenar hall, a viewing terrace accessible to the public, a circulation ramp that spirals upwards and the dome as a central design element. The Town Hall, which was occupied in 2002, is however overshadowed by the other buildings in the complex: the offices and hotels with their rather mundane glass architecture.

Restricted View?

It is still uncertain whether a further prestige project will be built in the immediate vicinity of London Bridge. Renzo Piano has designed a 305m high skyscraper for the Sellar Property Group as a symbol visible from afar of the revitalisation of this district. Although both the Mayor of London, Ken Livingstone, and Southwark Council are in favour of the project the decision was passed on to the Deputy Prime Minister on account of the national significance of the site. Organisations such as English Heritage fear for the quality of the historic surroundings and worry that visual connections to important witnesses of London's history such as St. Paul's Cathedral could be impaired.

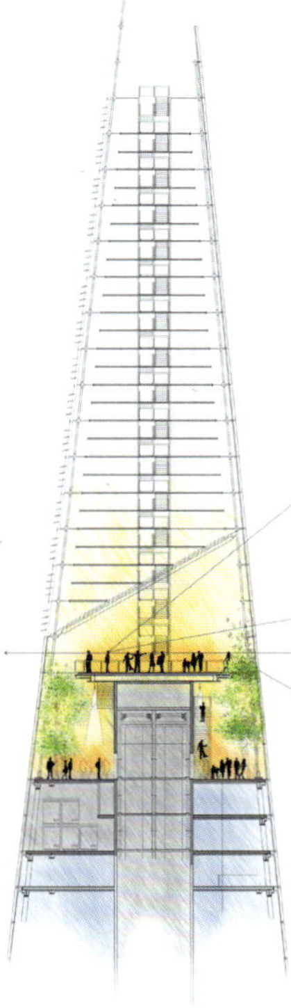

Renzo Piano: London Bridge Tower

Photos RPBW, Hayes Davidson/John Maclean, Frédéric Terreaux

Will Alsop: Palestra building

Photo Studio Alsop

fußgängerfreundliche Gestaltung sorgt die zentrale unterirdische Verkehrserschließung.
Ein besonderes Beispiel von Public-Private-Partnership verkörpert die Anmietung des privatwirtschaftlich errichteten Rathauses durch die Greater London Authority. Das eiförmige und völlig verglaste Gebäude beherbergt einen Versammlungssaal, Ausschussräume, Büros und ein Café. Wichtige Merkmale sind der runde Plenarsaal, eine öffentlich zugängliche Aussichtsplattform, eine sich spiralförmig in die Höhe schraubende Rampe zur Erschließung und die Kuppel als zentrales gestalterisches Element. Das 2002 bezogene Rathaus wird jedoch von den anderen Gebäuden des Komplexes, Büros und Hotels in eher konventioneller Glasarchitektur, in den Schatten gestellt.

Eingeschränkte Sicht?

Noch ist unklar, ob ein weiteres Prestigeprojekt in direkter Umgebung der London Bridge errichtet wird. Renzo Piano hat für die Sellar Property Group einen 305 m hohen Wolkenkratzer entworfen, als weithin sichtbares Zeichen der Revitalisierung des Gebietes. Obwohl der Bürgermeister Londons, Ken Livingstone, sowie das Southwark Council das Projekt befürworten, wurde die Entscheidung auf Grund der nationalen Bedeutung an den Deputy Prime Minister verwiesen. Organisationen wie English Heritage fürchten um die Qualität der historischen Umgebung und um die Einschränkung der Sichtbeziehungen zu wichtigen Zeugnissen der Geschichte Londons wie St. Paul's Cathedral.
Neben hochwertigen Büros sollen in dem Gebäude eine Reihe von öffentlich zugänglichen Aussichtsplattformen sowie Museen und Begegnungszentren entstehen, u.a. zwei „Piazzas in the sky". Auch die üblichen Bürofolgeeinrichtungen, wie Restaurants, Geschäfte und ein Fitnessclub, werden direkt im Gebäude untergebracht.
Nach einem Auftakt mit kulturellen Highlights nimmt so die Bedeutung des Südufers als Büroquartier zu. War es bis in die frühen 1990er Jahre Standort von Backoffices, ziehen inzwischen auch Headquarter großer Konzerne in das Gebiet, z.B. in das geplante „Palestra"-Gebäude an der U-Bahn Station Southwark, entworfen von Will Alsop. Internationalen Trends folgend, hat man sich jedoch von monofunktionalen Projekten verabschiedet, zumindest im Erdgeschoß werden meist Gastronomie und Einzelhandel untergebracht.
Mutige und besonders innovative Architektur findet man in der Zone am Fluss, die starken kommerziellen Verwertungsinteressen ausgesetzt ist, kaum. Auf einige Kleinodien trifft man erst in weiter abseits gelegenen Gebieten, z.B. auf das in leuchtendem Orange und Pink gehaltene Museum of Fashion and Textiles von Ricardo Legorreta in der Bermondsey Street. Für Planer und Politiker jedoch ist der Stadtumbau am Themseufer ein wahr gewordener Traum: Londoner und Touristen zugleich strömen in eine ehemalige no-go-area. Der an der Themse entstandene "Freizeitpark des gehobenen Geschmacks" steigert die Anziehungskraft Londons auf kulturinteressierte Städtetouristen. Da die neue Aufenthaltsqualität, gefördert durch schicke Restaurants und Geschäfte sowie pausentaugliche öffentliche Räume, von Besuchern und Angestellten wahrgenommen wird, steigt gleichzeitig die Attraktivität des Standortes für internationale Dienstleistungsunternehmen und ihre hochqualifizierten Arbeitskräfte. Der Status Londons als Tourismusmetropole und World City scheint also gesichert.

In addition to high-quality office space, a number of viewing terraces open to the public as well as museums and encounter centres are planned in the building, including two 'piazzas in the sky'. The usual service facilities associated with offices such as restaurants, shops and a fitness club are to be located directly in the building.
After an initial phase dominated by cultural highlights, the significance of the south bank as an office location is increasing. Whereas until the early 1990s it was primarily a location for back offices, in recent times headquarters of large companies have begun to move into the area, for example in the planned Palestra building at Southwark Tube Station, designed by Will Alsop. Following international trends, mono-functional projects have been abandoned and, at least at ground floor level, shops and restaurants are integrated.
Courageous and particularly innovative architecture is hardly to be found at all in the zone directly at the river, which is subject to strong commercial interests. Small jewels, such as the bright pink and orange Museum of Fashion and Textiles by Ricardo Legorreta in Bermondsey Street, can be discovered in the backwaters. However, for planners and politicians alike the urban renewal project on the Thames bank is a dream come true. Both Londoners and tourists stream into a former no-go area. The 'good taste leisure park' increases London's attractiveness as a destination for city tourist interested in culture. The new quality created by smart restaurants and shops as well as open spaces suitable for short lunchtime breaks means that the area is used to an equal extent by visitors and office employees. This in turn increases its attractiveness as a location for international service industry companies and their highly qualified workforces. London's status as tourist metropolis and world city thus seems assured.

Kop van Zuid

Kop van Zuid

Pragmatisches Panoptikum – Rotterdams Hafenumbau
A Pragmatic Panopticum – The Rebuilding Of Rotterdam's Port

Photos a+t, Rob't Hart, Christian Richters, Nigel Young, Hans Werlemann
Text Klaus-Dieter Weiß

Rotterdam: die Stadt am Fluss, die Stadt von Rem Koolhaas und MVRDV, die Stadt der Urbanität und des Optimismus', der wilden Kontraste, der schlagfertigen Experimente und kühnen Visionen, der 162 verschiedenen Nationalitäten ist ohne Beispiel. Bis 2010 soll der Sprung über die Maas gelingen.

Rotterdam: the city on the river, the city of Rem Koolhaas and of MVRDV, a city of urbanity and optimism, of wild contrast, off the cuff experiments and daring visions, of 162 different nationalities is a city that cannot be compared to any other. Now the goal there is to make the leap across the Maas by spring 2010.

Bauten von rechts nach links | buildings from right to left
Wiel Arets, UN Studio, Renzo Piano, Bolles + Wilson

Am wenigsten vergleichbar ist Meinhard von Gerkans bis 2005 aus dem Nichts entstehende, streng geometrisch konzipierte Idealstadt Luchao Harbour City - Brückenkopf für den 30 km weit auf Inseln vorgelagerten internationalen Container-Tiefseehafen Yangshan. Eine funktional gegliederte und formal per Stadtstilgutachten dekretierte Hafenstadt 50 km südwestlich von Shanghai, rund um den Fokus eines kreisrunden, „unberührbaren" Sees. Straßenbahnen, die das nur mit wenigen Kulturbauten und Freizeitangeboten besetzte Sonnensymbol von 2,5 Kilometern im Durchmesser in ewig gleicher Himmelslogik links- oder rechtsherum umrunden, machen den Sonnenkult perfekt. Kann eine Stadtneugründung für 300.000 Menschen, im Endausbau halb so groß wie Rotterdam, Urbanität wirklich als konzentrische Generalstabsplanung synthetisieren? Im radialen Querschnitt (See, Promenade, Businessdistrikt, Stadtpark, Wohnquartiere) zielen alle Straßen unerbittlich auf eine „Wolkennadel" in Seemitte. Quer dazu verbiegt sich alles zum Kreis. Mit dieser Hommage auf die Stadtidee von Sonnenkönig und Sonnenzar, die selbst Peter der Große für seine Neugründung St. Petersburg zugunsten des lebhafteren Vorbilds Amsterdam verwarf, wendet sich die chinesische Hafenstadt in unerbittlicher Konsequenz mit ihrer Rückseite zum Meer.

Das vor 720 Jahren erstmals erwähnte Rotterdam kämpft stattdessen um den städtebaulichen Sprung über die Maas, um jeden Einwohner, der im Zentrum wohnt, um jede Abweichung, um jede Akzentuierung, um jeden Wolkenkratzer. Um jeden Spleen? In den Niederlanden werden auch Kirchen, Schiffe und – immer noch intakt und beliebt – Piet Bloms 1984 inszenierte Operettenstadt der „Baumhäuser" bewohnt, warum nicht Hochhäuser? Mit den Häusern auf der gegenüber liegenden Seite des heutigen Museumshafens „Leuvehaven" entstanden 1978 zum ersten Mal im Zentrum von Rotterdam wieder Wohnungen. Die Nachkriegsstadt, kaum älter als die darin zu findende erste Fußgängerzone Europas von 1951, lebt von ihrer Dichte und Komplexität. Man wird dem Manhattan an der Maas, dem soviel amerikanisches Chaos wie deutsche Disharmonie nachgesagt wird, kaum vorwerfen können, in dieser kurzen Zeit nicht zu sich selbst gefunden zu haben.

Least comparable to Rotterdam is Meinhard von Gerkan's geometrically conceived ideal city, Luchao Harbour City, to be stamped out of nothing by 2005. A bridgehead for Yangshan international container deep-sea port located on islands 30 kilometres away. A functionally organised port, 50km south west of Shanghai, formally ordained in decree on urban style, built around the focal point of a circular "untouchable" lake. Trams circling around the symbol of the sun, about 2.5km in diameter and occupied by only a few cultural buildings and leisure time facilities, always moving in the same direction – either clockwise or anticlockwise, making the sun cult perfect. Can a new city for 300,000 people, in its final planned extent half as large as Rotterdam, really synthesise urbanity in the form of concentrated general command-type planning? In the radial cross-section (lake, promenade, business district, town park, residential districts) all streets are oriented remorselessly towards a Wolkennadel (Cloud Needle) in the middle of the lake. At right angles to these radial streets everything revolves in circles. With this homage to the urban planning ideas of the Sun King and Sun Czar, which even Peter the Great rejected in his plans for the new city of St Petersburg in favour of the livelier example of Amsterdam, the Chinese port turns its back to the sea with a merciless consistency.

Rotterdam, on the other hand, first mentioned in documents 720 years ago, is fighting about the urban leap over the Maas, about every inhabitant that lives in the centre, about every difference, about every accentuation, about every skyscraper. About every spleen? In the Netherlands churches, ships and Piet Blom's operetta town of 'tree houses' – built in 1984 and still intact and popular – are lived in, why not high rise buildings? With the buildings erected in 1978 on the far side of what is today the museum port of Leuhaven, apartments were built again for the first time in central Rotterdam. The post-war city, little older than the first pedestrian zone in Europe dating from 1951 and also to be found in Rotterdam, lives from its density and complexity. One can hardly accuse the Manhattan on the Maas, which is said to have as much American chaos as it does German disharmony, of not having found itself within this short period.

Bolles + Wilson: Uferanlagen
Embankment

UN Studio: Erasmus Brücke
Erasmus bridge

Avantgarde-City

Die Maas mit Innenstadtqualitäten zu überspringen, um die seit 1874 bestehende Teilung von City diesseits und Hafen jenseits aufzuheben, war für die Stadtentwicklung von entscheidender Bedeutung. Adolf Max Vogt sieht den wesentlichen Vorzug von Wasserstädten darin, mit dem Wasser Raum und Atem zu gewinnen für kritische Innenansicht und Selbstbetrachtung. Gerade weil der Stadt der Zugang zum Wasser lange versperrt war, sind heute alle Chancen einer intensiven Beziehung gegeben. Doch gerade die entscheidende Wasserfront Rotterdams, die Uferstraße Boompjes gegenüber der Maasinsel Noordereiland, ist städtebaulich noch nicht sehr gesprächig, trotz der 1989 von OMA neu gestalteten Kaianlage. Nach zahlreichen Neuansätzen und Workshops beginnt die Revitalisierung der viel zu kargen Schauseite mit ihren 200.000 Binnenschiffen pro Jahr erst jetzt mit einer ersten Baustelle. Altmeister Philip Johnson baut auf Höhe des Uferrestaurants von Mecanoo (1990) für die Firma Ernst & Young einen 78 m hohen Büroturm. Spannender ist die zweite Phase dieses Projekts, ein ebenso hoher Turm mit Luxus-Apartments (Architekten: Brouwer Steketee, Baubeginn: 2004), der sich aus der zweiten Reihe wie ein Zaungast über einen sechsgeschoßigen Altbau der Nederlandsche Bank reckt.

Das Vertrauen in internationale Architekturgrößen ist in Rotterdam oft genug enttäuscht worden. Trotzdem ist diese Sammelleidenschaft Programm, als ginge es seit dem Hochhaus der Generale Bank 1996 von Helmut Jahn um eine Musterausstellung des Architekturmuseums. Der in Toronto kreierte Hotel- und Büroturm Millennium in Bahnhofsnähe zum Beispiel (Architekten: Webb Zerafa Menkes Housden, Baujahr: 2000) lässt eher auf einen Höhepunkt der Postmoderne schließen als auf einen architektonischen Aufbruch ins neue Jahrtausend. An die Qualität des Hochhausensembles, das Rem Koolhaas und Kees Christiaanse 1980 für die Uferstraße Boompjes entwarfen, scheint sich heute niemand mehr erinnern zu wollen. Entgegen der üblichen niederländischen Verwaltungspraxis, einen kleinen Kreis von anerkannt innovativen, erfahrenen wie jungen Architekten undemokratisch zu bevorzugen, beugt sich ausgerechnet Rotterdam, der nur Hamburg vergleichbare Wegbereiter einer vorbildlichen Architekturkultur, allzu oft dem Diktat der Investoren und ihrer inszenierten Sachzwänge.

Von den Distanzen der City selten überforderte Fußgänger erreichen zwischen Landstadt und Wasserstadt dennoch mindestens die Erlebnisdichte eines Autofahrers in Berlin. Zwischen Bahnhof und Erasmus-Brücke liegen kaum 2.000 Meter. Trotzdem ist dieser scheinbar planlose labyrinthische Urbanismus ohne populäre Sehenswürdigkeiten und nur in einem Stadtteil historisch kein anerkanntes touristisches Ziel. Der Delfshaven, ein halbes Jahrtausend der über einen Kanal angebundene Hafen von Delft, gehört erst seit 1886 zu Rotterdam. In der überbordenden Reiseliteratur erscheint die zweitgrößte niederländische Stadt lediglich als Randnotiz. Den Ursachen nachzuspüren könnte der Stadtentwicklung eine neue Richtung geben. Das seit den 1970er Jahren geplante Vorzeigeviertel Kop van Zuid, der ehemalige Polder „Feyenoord", den schon die ersten Erweiterungspläne der Stadt im Jahr 1858 im Blick hatten, lässt einen derartigen Richtungswandel nicht erkennen, obwohl die touristische Magnetkraft ausdrückliches Ziel war.

Avant-Garde City

To leap over the Maas employing inner urban qualities, to eliminate a division existing since 1874 between the city on one side and the port on the other, was of decisive importance for urban development. Adolf Max Vogt sees the primary advantage of cities on the water in the fact that, due to the water, space and breath can be drawn for critical internal examination and self observation. Precisely because the city was long denied access to the water, today all possible chances for an intensive relationship exist. But the decisive waterfront of Rotterdam, the waterside street Boompjes, opposite the Maas island Noordereiland, is not particularly loquacious in urban planning terms, despite quays newly designed by OMA in 1989. After numerous new starts and workshops the revitalisation of the extremely bare show face with its 200,000 vessels per year is only now beginning with the first building site. Old master Philip Johnson is building a 78m high office tower for Ernst & Young in the area of the waterside restaurant by Mecanoo (1990). More exciting is the second phase of this project, a tower of the same height with luxury apartments (architects: Brouwer Steketee; start of construction: 2004), which peeps like a voyeur from the second row over an old six-storey building of the Nederlandsche Bank.

The faith in internationally renowned architects' names has often been disappointed in Rotterdam. Nevertheless, this passion for collecting famous signatures remains a programme, as if ever since Helmut Jahn's 1996 high-rise building for the Generale Bank, the idea had been to create a model exhibition like in an architecture museum. The Millenium hotel and office tower close to the railway station, created in Toronto (architects: Webb Zerafa Menkes Housden; construction 2000), is more reminiscent of a highpoint of post-modernism than of an architecture emerging into a new millennium. Nobody appears to remember the quality of a high-rise ensemble designed by Rem Koolhaas and Kees Christiaanse in 1980 for the waterfront street Boompjes any longer. In contrast to the usual Dutch practice of undemocratically preferring a small circle of recognised, innovative and experienced young architects, Rotterdam of all places, comparable only with Hamburg as the precursor of an exemplary architectural culture, is yielding too often to the dictates of the investors and their dramatically presented constraints.

Pedestrians, who are seldom overtaxed by the distances in this city, experience a density of experiences between the land-town and the water-town that is at least the equivalent of that of a motorist in Berlin. The distance between the railway station and the Erasmus Bridge is scarcely 2000 metres. Nevertheless this apparently unplanned labyrinthine urbanism, without popular sights and only in one part historic, is not a recognised tourist goal. Delfshaven, the port of Delft connected for over five hundred years by a canal, has only belonged to Rotterdam since 1886. In the overwhelming amount of travel literature the Netherlands' second-largest city appears only as note in the margins. Tracing the reasons for this could point urban development in a new direction. The model district of Kop van Zuid, on the drawing board since the 1970s, formerly the polder Feyenoord, which the first expansion plans for the city in 1858 had in its sights, does not suggest any change in direction, although tourism as a magnet was a clearly expressed goal.

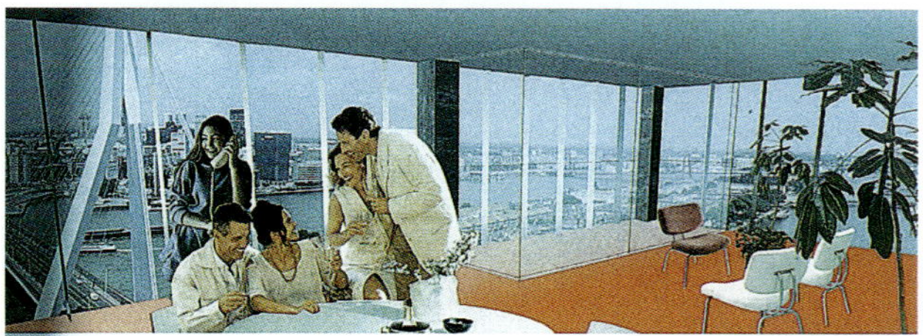

OMA/Rem Koolhaas:
„De Rotterdam", Wilhelmina-Pier

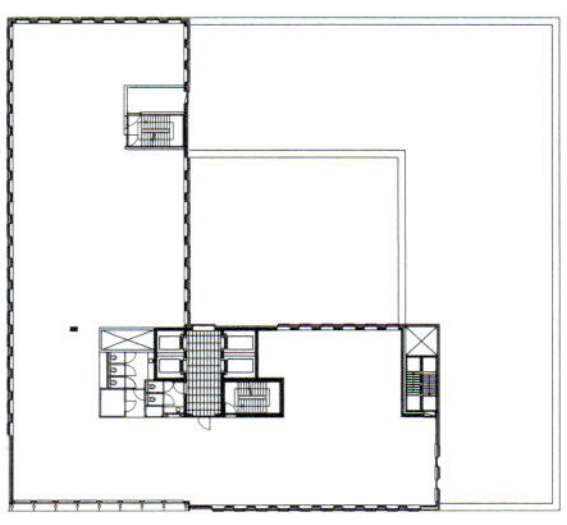

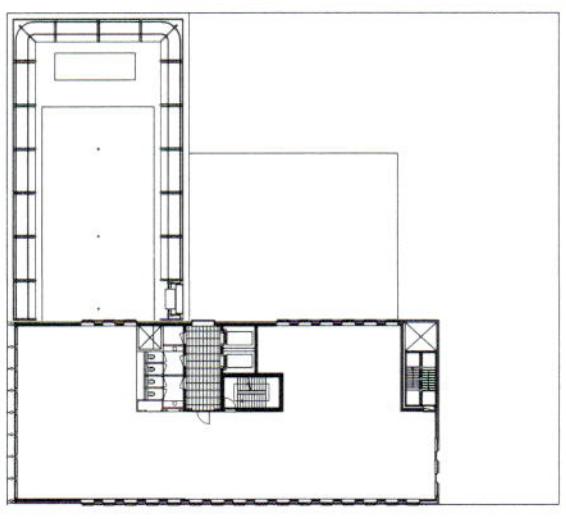

Meyer & van Schooten:
„Head Office", Bürogebäude, Zuidkade
Office building

Turbomoderne

„High-rise is a Must!“[1] und „Accelerating Rotterdam“[2] lauten die Losungen einer Stadt, in der zum Jahrtausendaufbruch eine umtriebige „Skyscraper-Foundation“ ins Leben gerufen wurde (die neuesten Projektentwicklungen, Arbeitsmodelle und Baufortschritte meldet www.wolkenkrabbers.nl). So überrascht es nicht, dass Rotterdam in den Wolkenkratzer-Disziplinen Rathaus, Restaurant, Leuchtturm, Brückenpylon, Schornstein und Wohngebäude alle niederländischen Konkurrenten hinter sich lässt. Vermutlich wird der Wettkampf um Höhenmeter hollandweit aber auch nur innerhalb der von Rem Koolhaas 1978 ausgerufenen und von MVRDV in zweiter Generation betriebenen „Kultur der Dichte“ so ernst genommen. Am beliebtesten scheinen die Fotoperspektiven der Stadt, die dem amerikanischen Idealbild mit starken Teleobjektiven am nächsten kommen. Das Himmelstürmen gilt als Symbol für Modernität und Urbanität, Innovation und Beweglichkeit, Zukunftsfähigkeit und Internationalität. Niemand ereifert sich darum, wenn Mecanoo aus dem Hinterhalt einer Bebauung des 19. Jahrhunderts einem Kirchturm mit einem dynamischen Hotelturm in die Quere kommt (Park Hotel, 1992). Das magische Jahr 2000 wurde mit dem Entwurf „Tower 2000“ schon 1970 in den Blick genommen. Eine gigantische Blumenvase nach dem Vorbild des Eiffelturms, jedoch exakt 2.000 Meter hoch. Am liebsten würde man die mit über zwei Kilometern längste Einkaufsstraße Rotterdams, Nieuwe Binnenweg, als Wolkenkratzer senkrecht stellen. In Abwandlung einer Idee, die MVRDV mit seinem Europan-Beitrag von 1991 für einen Berliner Baublock im Sinne hatte. Soviel heroische Moderne war seit den raffinerieähnlichen Stadtvisionen von Archigram selten auf Papier zu bestaunen. Die realisierte architektonische Qualität ist allerdings zu diskutieren. Immerhin konzentriert sich der Hochhaus-Boom auf klar definierte Achsen diesseits und jenseits der Erasmus-Brücke, deren Nord-Südausrichtung einen Stadtboulevard definieren soll. Noch vor drei Jahren wurden die Zwillingstürme des New Yorker World Trade Center als typologisch gleichwertig in eine Reihe mit Exemplaren in San Gimignano und Rotterdam gestellt: den 1999 realisierten, 102 Meter hohen Doppel-Wohntürmen „Hoge Heren“ von Wiel Arets am cityseitigen Fußpunkt der Erasmus-Brücke. Leider stören die schwarz gekleideten „Herren“, aus der südlichen Sichtachse betrachtet, empfindlich die Skyline der Brücke. Stolz kann Rotterdam die Himmelstürmerei auf das Jahr 1898 zurückverfolgen. Der nur 45 Meter hohe Massivbau „Het Witte Huis“ in Ufernähe der Maas, einer der wenigen populären historischen Fixpunkte der Stadt, war lange Zeit Europas höchstes Bürohaus. Aber kann das ehemalige Hafengelände Kop van Zuid jenseits Ben van Berkels großer Eröffnungsgeste, der Erasmus-Brücke aus dem Jahr 1996, den Höhenschwur tatsächlich halten? Bisher nicht, leider nicht. Die Türme Norman Fosters für das World Port Center und Renzo Pianos für die Telecom, standesgemäß mit einer 1.000-Pixel-Nachrichten-Fassade ausgestattet, sind nicht nur nach Höhenmetern, sondern auch architektonisch keine Meilensteine. Doch in unmittelbarer Nachbarschaft ist Abhilfe jetzt in spektakuläreren Dimensionen mit Projekten von Rem Koolhaas und Mecanoo in Sicht. Mit spielerisch variierten Fassaden und Bauhöhen bis zu 150 Metern dokumentiert Mecanoo die Nutzungsmischung des gerade begonnenen Projekts Montevideo. Neben 192 exklusiven Apartments zwischen 253.000 Euro für durchschnittlich 127 Quadratmeter und 2.250.000 Euro für das Penthouse mit Pool sind 6.000 Quadratmeter für Büros, knapp 2.000 Quadratmeter für Läden und ein Fitness-Club vorgesehen. Mit seinem immer wieder verschobenen, 135 Meter hohen Thesenbau „De Rot-

Turbo-Modernity

High-rise is a Must![1] and Accelerating Rotterdam[2] are the names of the solutions in a city in which, at the beginning of the millennium, a high profile Skyscraper Foundation was called into being (the latest project developments, working models and the state of building progress can be viewed at www.wolkenkrabbers.nl). So it is not particularly surprising that in the skyscraper disciplines of town hall, lighthouse, bridge pylon, chimney-stack, and apartment building Rotterdam leaves all other Dutch competitors far behind. It is likely that the competition about heights in Holland is only taken seriously within the 'culture of density' called into being by Rem Koolhas in 1978 and promoted in the second generation by MVRDV. The most popular items are the photo-perspectives of the city that, using powerful telescopic lens, come closest to the American ideal image. This conquest of the skies is seen as a symbol of modernity, urbanity, innovation, flexibility, forward-thinking and internationalism. Nobody worries much when out of the backyards of a 19th century development a dynamic hotel tower by Mecanoo infringes on a church steeple (Park Hotel, 1992). The magical year 2000 was already focused on back in 1970 in the design Tower 2000. A huge flower vase based on the Eiffel tower but exactly 2000 metres high. What many would most like to do would be to erect Rotterdam's longest shopping street Nieuwe Binnenweg, more than two kilometres long, vertically as a skyscraper. A transformation of an idea that MVRDV used in their Europan contribution in 1991 with a Berlin building block. Seldom, since the refinery-like urban visions of Archigram, has there been so much heroic modernism to admire on paper.

But the quality of the built architecture is open to discussion. Whatever the case the skyscraper boom is concentrated on clearly defined axes on the near and far sides of the Erasmus Bridge, whose north-south orientation is intended to define a city boulevard. Only three years ago the twin towers of the New York World Trade Center were placed on the same typological level as towers in San Gimigniano and in Rotterdam: the 102 metre high double apartment towers, Hoge Heren, built in 1999 by Wiel Arets at the city-side base of the Erasmus Bridge. Unfortunately, the "gentlemen dressed in black" distinctly disturb the skyline of the bridge when viewed along the southern axis. Proudly Rotterdam can trace its tradition of building into the heavens back to 1898. The only 45m high massive Het Witte Huis, close to the banks of the Maas and one of the city's few popular historical fixed points, was for a long time Europe's highest office building – but can the former port Kop van Zuid on the far side of Ben van Berkel's expansive opening gesture, the Erasmus Bridge from 1996, in fact keep the vow of height? Up to the present, unfortunately not. The towers by Norman Foster for the World Port Center and Renzo Piano for Telecom, fittingly equipped with a 1000 pixel news facade, are far from architectural milestones in terms of height and architectural quality. But in the immediate neighbourhood there is help at hand in the form of projects with spectacular dimensions by Rem Koolhaas and Mecanoo. With playfully varied facades and building heights of up to 150 metres, Mecanoo documents the mix of use of the recently started Montevideo project. In addition to 192 exclusive apartments between 253,000 Euro for, on average, 127 square metres and 2,250,000 Euro for the penthouse with pool, 6000 square metres of office space, almost 2000 square metres for shops and a fitness club are planned. With his repeatedly postponed 135 metre high thesis building De Rotterdam, Rem Koolhaas hopes to realise his old dream of

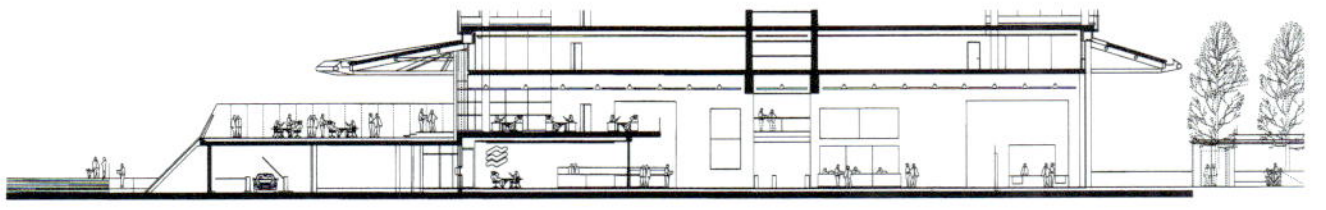

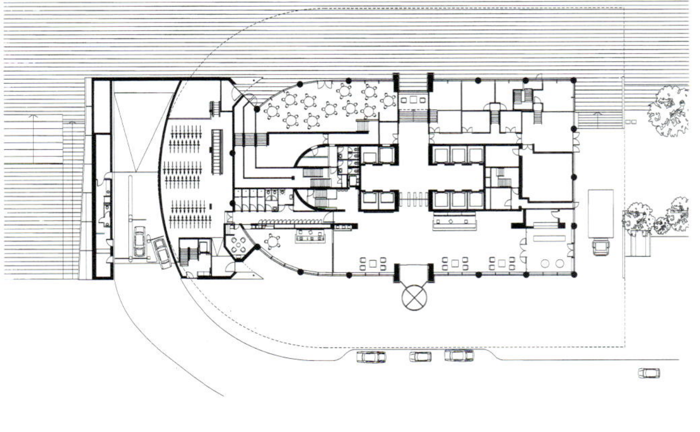

Norman Foster:
World Port Center

terdam" möchte Rem Koolhaas seinen alten Traum der „vertikalen Stadt" realisieren: Einkaufszentrum, Büros, 260 Apartments, Parks, Kinos, Hotel, Gastronomie, Fitness, 600 Parkplätze in einem Mix privater und öffentlicher Flächen. Baubeginn soll dieses Jahr sein, doch die Nachfrage scheint weiterhin fraglich.

Giebelhausromantik

Pim Fortuyn, der kurz vor seiner Umkehrung der politischen und architektonischen Verhältnisse ermordete Rotterdamer Soziologe und Präsidentschaftskandidat, schmiedete in der Zerrissenheit der Stadt zwar eine Allianz aus Postmoderne und Rechtspopulismus, wollte sich in seiner populären Kritik an den Neubauplänen auf Kop van Zuid aber nicht auf Alternativen festlegen. Bisher wurde die Berlin ähnliche Zerrissenheit Rotterdams offensiv thematisiert, nicht durch eine sogenannte kritische, morphologische Rekonstruktion, die sich historisch verziert längst in eine populistische verkehrt. Für diese Verirrung sorgt in Rotterdams Vorzeigeviertel Hans Kollhoff, mit seiner gerade begonnenen Giebel-Offensive „De Compagnie". Schon vor hundert Jahren galt es dem aufstrebenden Amerika als chic, Gefühle kultureller Unterlegenheit eklektizistisch abzuschütteln. Neureiche Magnaten spielten den Renaissance-Fürsten und verlangten nach Bühnen im Beaux-Arts-Stil. Nicht nach ahistorischem Modernismus im Gefolge von Gewächshäusern und Ausstellungshallen. Heute grassiert mit den uneinnehmbaren Märchenschlössern des Separatismus eine neue Vergangenheitswut. Angeblich Verbindliches und Unverfälschbares steht hoch im Kurs, es lockt das stets vergangene Glück im Gewesenen. Hans Kollhoff schürt dieses Fieber mit gotischen Kirchtürmen, goldenen Zinnen, Anleihen bei Schinkel, Ledoux und Manhattan.
Dieses letzte Missverständnis muss ihm wohl die Tür nach Rotterdam geöffnet haben. Architektonische Fluchthelfer wie er scheitern jedoch daran, dass Komplexität und Dichte des Originals mit den heutigen Mitteln und Fähigkeiten nicht reproduzierbar sind.

Volksbelustigung

Hans Kollhoff scheint sich dieser Illusion bewusst zu werden, wenn er auf Kop van Zuid in einen Rationalismus nach Aldo Rossi verfällt: 16 Geschoße Rasterfassade plus steiles Satteldach und Gaupen in Doppelreihe. Der Traum von der stillgestellten Architektur mag noch so verlocken, das Schicksal dieses kurzfristigen Publikumserfolgs liegt in seiner Nähe zu Walt Disney. Ohne deutlichen Gegenwartsbezug ist Retro nicht entwicklungsfähig. Der Brückenschlag in die Vergangenheit oder vielmehr aus ihr heraus gelingt nur dann, wenn das gedankliche Tragwerk nicht reine Nostalgie bleibt. Den hilflosen Glauben an das Neue durch einen noch hilfloseren Glauben an das Alte auszutauschen bleibt frommer Wunsch, die Realisierung so fade wie fadenscheinig. Echte Vergangenheit nimmt Schaden. Warum über eine falschgestellte Zeit nachdenken? Wie endete Gesine Weinmiller, Kronzeugin und ehemalige Projektleiterin aus dem Büro Kollhoff, in ihrem bemerkenswerten Vortrag „Der Traum von der Klassik"?[3] „Ein Dieb wird auch nicht durch das Tragen eines Zylinders zum Grandseigneur." Trotzdem steht auch Rotterdams Innenstadtviertel am Südufer der Maas nach den erasmischen Glashäusern einer offenen, geheimnislosen Gesellschaft[4] vor der Wende vom Supermodernismus zu einem Konservatismus, dem neupostmoderne, individualistische Spielereien von Kollhoff oder Krier wichtiger sind als der liberale Austausch zwischen Öffentlichkeit und Privatsphäre. Aufgrund eines Volksentscheids entschloss sich die Stadt Groningen, die gesamte Innenstadt im Stil des 19. Jahrhunderts

the vertical city: shopping centre, offices, 260 apartments, parks, cinemas, hotel, restaurants, fitness centre and 600 car parking spaces in a mix of private and public areas. Construction is due to commence this year but the demand still seems doubtful.

The Romanticism Of The Gabled Building

Pim Fortuyn, Rotterdam sociologist and presidential candidate, murdered shortly before his reversal of the political and architectural situation, was able to forge an alliance between postmodernism and right-wing populism, although he did not wish to commit himself to any alternatives in his populist criticism of the new construction plan for Kop van Zuid. Previously the fragmented quality of Rotterdam, similar to Berlin, was addressed directly, not by means of a so-called 'critical and morphological reconstruction' which, with its historicist ornament, has long since moved in a populist direction. This kind of confusion was introduced to Rotterdam's model district by Hans Kollhoff with his recently started gable offensive De Compagnie. One hundred years ago in upwardly mobile America it was regarded as chic to eclectically shake off feelings of cultural inferiority. Nouveau-riche magnates played the Renaissance prince and demanded backdrops in Beaux Arts style. And not non-historical modernism accompanied by greenhouses and exhibition halls. Today a new kind of diseased obsession with the past is flourishing in the form of impeachable fairy tale castles of Separatism. That which supposedly connects with our past and cannot be falsified experiences a boom, the (always past) happiness of days gone by exerts its seductive power. Hans Kollhoff intensifies this fever with gothic steeples, golden pinnacles and elements borrowed from Schinkel, Ledoux, Manhattan. It must have been this latter misunderstanding that opened the door to Rotterdam for him. But architectural flying doctors such as Kollhoff are always doomed to failure, as the complexity and density of the original cannot be reproduced with the means and capabilities available today.

Popular Entertainment

Hans Kollhoff appears aware of this illusion when he succumbs to a kind of rationalism á la Aldo Rossi in the case of Kop van Zuid: a 16 storey gridded facade plus a steep pitched roof and two rows of dormer windows. The dream of a pacified architecture may still be seductive but the destiny of this short-term public success lies in its closeness to Walt Disney. Without clear references to the present, retro is incapable of further development. Building a bridge to the past or, to put it more accurately, from the past succeeds only when the intellectual substructure is not based entirely on nostalgia. Hoping to exchange a helpless belief in the new for an even more helpless belief in the old remains a pious wish – the realisation is as boring as it is specious. The true past is damaged as a result. Why reflect about an inaccurately perceived time? How did Gesine Weinmiller, crown witness and former project director in Kollhoff's studio, end her remarkable lecture Der Traum Von Der Klassik (The Dream Of The Classical)?[3] "A thief does not become a grand gentleman by simply wearing a top hat." Nevertheless Rotterdam's inner city district on the south bank of the Maas, following the Erasmian glasshouses of an open society, free of secrets,[4] is confronted with a transition from supermodernism to a kind of conservatism in which the new post-modern individualist games played by Kollhoff or Krier are more important than a liberal exchange between the public and private realms. On the basis of a referendum, the town of Groningen has decided to reconstruct the entire inner city in the

Bolles + Wilson: Luxor Theater

zu rekonstruieren. „In Rotterdam, der früheren Basis der Avantgarde, gehört es längst zum guten Ton, Bauwerke wie die Kunsthalle von Rem Koolhaas unmöglich zu finden und ihren Abriss zu fordern."[5] Der große Humanist Desiderius Erasmus von Rotterdam, bekanntester Sohn der Stadt, verliert damit architektonisch an Glanz. Im Vorzeigestaat der ewigen Moderne und Demokratie geraten seine Jünger trotz des gerade neu aufgelegten neoerasmischen Manifests „Supermodernismus" des am Rotterdamer Architekturmuseum wirkenden Kunsthistorikers Hans Ibelings in Erklärungsnot.[6] Zwar kommt die Postmoderne auf Kop van Zuid neben Kollhoffs Beitrag formal nur in Doppelhäusern rund um den W.G. Witteveenplein zum Ausdruck, ein Beitrag der Architekten Molenaar & van Winden ganz im Sinne sehr früher französischer Parallelen. Auch in sehr biederen und banalen Reihenhäusern oder Mietskasernen. Aber den Rückzug in die Festung von Wohngemeinschaft, Tennisplätzen und Privatwohnung dokumentieren sogar die hermetisch verriegelten terrassierten Wohnburgen der Architektengruppe „de Architekten Cie."[7] Die Öffnung zum Wasser, das grandiose Thema der niederländischen Grachtenstadt, unterbleibt zugunsten der privaten Sicherheit. Die in Straßenraum oder Platz kasernierte Öffentlichkeit bleibt damit leeres Versprechen.

Zu Beginn der Postmoderne war schon Piet Blom der Meinung, Städtebau habe der Volksbelustigung zu dienen; Wohnviertel in der Innenstadt seien nur dann human, wenn sie unterhaltsam wirken; Städtebau müsse Zirkus sein. Diesen Publikumswunsch erfüllt vor allem das Altstadt-Dreieck mit dem alten Hafen der Stadt. Der „Südkopf" Rotterdams jenseits der Maas ist dagegen ein sprödes Laboratorium für Architekturexperten, selbst die darin synthetisierte Postmoderne hat ihren Frohsinn verloren. Das früheste Spektakulum, die einer Büroklammer nachempfundene Wohnschlange „De Peperklip" von Carel Weeber aus dem Jahr 1982 wurde 1994 sogar in die Runde der geschmacklosesten Bauwerke des Landes gewählt.

Stadt von morgen

Wie die Städte des Ruhrgebiets ist die „große Werkstatt" Rotterdam eine typische Stadt des 19. Jahrhunderts. Ihr seit 40 Jahren weltweit größter Hafen erwirtschaftet 11 Prozent des niederländischen Bruttosozialprodukts und sichert 360.000 Arbeitsplätze. Mit allen ökologischen Folgen, die Rotterdam noch vor der Ersten Ölkrise erreichten. Mit allen gesellschaftlichen Problemen, die Anfang der 1980er Jahre in „Türkenkriegen" gipfelten. Bis Mitte der 1980er Jahre dauerte der Exodus der Stadtbevölkerung an, erst dann wurde das Sinnbild einer hellen, modernen und zukunftsorientierten Stadt mit Hilfe einer Architektur der „gastlichen Räume" reformiert. Erst seit 1986, mit dem politischen Programm „Das neue Rotterdam", wird der langfristige Einfluss von Umstrukturierungsplänen auf die Entwicklung des Stadtgebiets berücksichtigt. Seitdem werden Pläne für einzelne Stadtgebiete auf ihren Einfluss auf die Gesamtstadt geprüft. Noch heute überrascht die Präsenz von berittener Polizei in den niederländischen Innenstädten. Da in den Niederlanden positive regionale Ungleichgewichte durch fehlende lokale Gewerbesteuern ausgeschlossen sind, war Rotterdam an der Hafenwirtschaft nur negativ beteiligt. Wie der 93-jährige Lokalhistoriker Willem van Ravesteyn schilderte, schlug die zerstörerische Stunde der Moderne in Rotterdam dank rigoroser Industrialisierung und Hafenwirtschaft darum schon vor dem Zweiten Weltkrieg flächendeckend und vernichtend: „Die große Zerstörung von 1940, die übrigens soviel Hässlichkeit verschont hat, war eigentlich nur die Apotheose eines Vorgangs, der sich zeit meines Lebens mit voller Wucht

style of the 19th century. "In Rotterdam, previously a basis of the avant-garde, it is now considered civilized behaviour to find buildings such as the Kunsthalle by Rem Koolhaas impossible and to call for their demolition."[5] The great humanist Desiderius Erasmus of Rotterdam the city's most famous son, loses some of his architectural lustre in this context. In the model city of permanent modernism and democracy, despite the recently newly distributed neo-Erasmian manifesto Supermodernism by art historian Hans Ibeling, who works at the Rotterdam architecture museum, the followers of Erasmus are increasingly in need of elucidation.[6] Although on Kop van Zuid, apart from Kollhoff's contribution, post-modernism is only formally expressed by the double buildings around the W.G. Witteveenplein, a contribution by the architects Molenaar & van Winden entirely in the manner of early French equivalents, and in overly well-behaved and banal terrace housing or tenement blocks. But the retreat into the fortress of the housing community, tennis courts and the private apartment is documented even by the hermetically sealed terraced housing castles by the architects group de Architekten Cie.[7] The possibility of opening onto the water, the grandiose theme of the Dutch canal city, are ignored in favour of private security. The promise of a public realm interned in the street space or open squares remains an empty one.

At the beginning of post-modernism Piet Blom believed that the purpose of urban design was to keep the populace amused. He maintained that housing in the inner city is only human when it entertains; town planning should be a kind of circus. The triangle of the old town with the old port fulfils this function. The Südkopf of Rotterdam on the far side of the Maas is, in contrast, a sterile laboratory for architecture experiments – even the synthesised post-modernism there has lost its sense of fun. The earliest spectacle, a housing block shaped like a paper clip and appropriately named De Peperklip built by Carel Weeber in 1982 even joined the list of the most tasteless buildings in the country in 1994.

The City Of Tomorrow

Like the towns in the Ruhr district the 'major workshop' Rotterdam is a typical creation of the 19th century. Its port, for forty years the largest in the world, earns eleven percent of the Dutch gross social product and secures 360,000 jobs. With all the ecological consequences, which hit Rotterdam before the first oil crisis, also with all the social problems that peaked at the beginning of the 1980s with the 'Turkish Wars'. The exodus of the urban population continued until the mid 1980s, only then was the image reformed to create a bright, modern and future oriented city with the architecture of 'guest spaces'. It is only since 1986, with the political agenda for "the new Rotterdam" that the long-term influence of restructuring plans on the development of the urban area has been taken into account.

Since then plans for individual urban area are checked in terms of their influence on the development of the urban area as a whole. Today the presence of mounted police in the Dutch inner cities still remains surprising. As in the Netherlands positive regional imbalances are excluded by the lack of local industrial taxes, Rotterdam was involved in the port economy only in a negative sense. As the 93 year old local historian Willem van Ravesteyn describes, thanks to the rigorous industrialisation and port economy the destructive hour of modernism struck before the Second World War in a way that was extensive and absolutely destructive: "The great destruction of 1940 which, by the way, left so much ugliness intact, was in fact only the apotheosis of a process

Kruisheer/Elffers, Peter Wilson:
Albeda College, Parkstad

durchsetzte, aber eigentlich schon früher in der zweiten Hälfte des 19. Jahrhunderts begann."
In den ersten Kriegstagen wurde der historische Stadtkern bis auf Rathaus, Hauptpostamt, Börse und „Laurenskerk" ohne Luftalarm nahezu völlig zerstört – eine tragische Verkettung von Nachlässigkeiten und Missverständnissen nach einem deutschen Ultimatum. Nach Kriegsende wurden alle Grundstücke der Ruinenzone auf einen Schlag enteignet, die Straßen und sämtliche Infrastruktur sogar unterirdisch entfernt: freie Bahn für eine Reißbrettplanung, deren synthetischer „Städtebau" erst 30 Jahre später als gescheitert galt.
Vor diesem Hintergrund musste das Ziel der Stadt auf der Maas-Halbinsel Kop van Zuid darin liegen, hochgradige Funktionalität in innovativer Architektur zu realisieren. Nachdem Rotterdam jahrzehntelang als die „Neue Stadt" par exellence galt, wird ihre Erweiterung wiederum ganz unerschrocken als „City of Tomorrow" vorgestellt. Doch das Ergebnis ist abgesehen von einzelnen Bauten von Kees Christiaanse, Erick van Egeraat, Cepezed BV, Kruisheer/Elffers mit Peter Wilson und der sehr kohärenten, über mehrere kleine und große Projekte reichenden Mitarbeit von Bolles + Wilson ernüchternd. Die große Geste der Erasmus-Brücke ist noch keineswegs eingelöst, bis zum Endausbau des Jahres 2010 bleibt nicht viel Spielraum. In bester, wiederum freigeräumter Stadtlage, noch dazu allseits von Wasser umgeben und durchzogen, bleibt der Städtebau schematisch, unräumlich und unnahbar, die Architektur in den meisten Fällen uninspiriert. Nach den schlechten Erfahrungen auf dem ehemaligen Hafengelände der westlich benachbarten Halbinsel Katendrecht, wo noch Mitte der 1980er Jahre nach der Methode Tabula rasa Hafenbecken zugeschüttet wurden, um ein mono-funktionales Wohnviertel zu schaffen, für das der Fluss wiederum lediglich Stadtrand bzw. Hinterhof ist, verwundert das. Ist die Begeisterung für eine Stadt am Wasser doch durchaus im Publikum verankert und in zahllosen Beispielen positiv zu belegen. Nur sehr knapp konnte verhindert werden, dass auch auf Kop van Zuid die Kanäle zugeschüttet wurden. Unter der Leitung der für ihren Einsatz bereits mit dem Maaskantpreis ausgezeichneten Landschaftsarchitektin und Stadtplanerin Riek Bakker (Bakker & Bleeker), die seit dem entscheidenden Jahr 1986 auch das Stadtplanungsamt leitete, wurde am 27. August 1991 sogar eine Qualitätskontrolle mit weitreichenden Befugnissen auf sämtlichen Planungsebenen eingerichtet. Diesem Quality-Team gehört neben niederländischen Fachleuten, einem Spanier und einem Schweden unter anderem Mike Guyer (Gigon/Guyer) an.
Mit der Verlegung der Hafenaktivitäten zum Europort verkamen die verbliebenen Wohnviertel der Hafenarbeiter durch hohe Arbeitslosigkeit und Kriminalität zum Sanierungsgebiet. Konkret wurde die Planung von Kop van Zuid mit einem Architektursymposium im Jahr 1982, das bezeichnenderweise die Rotterdamer Kunststiftung organisierte, um auf diese Weise die architektonische Eintönigkeit der ersten Realisierungen bloßzustellen. Den beteiligten Architekten Aldo Rossi, Josef Paul Kleihues, Oswald Mathias Ungers und Derek Walter war erstmalig die Aufgabe gestellt, das gesamte alte Hafengelände am Südufer der Maas zu planen. 1987 ließ sich Teun Koolhaas, der Neffe von Rem Koolhaas, bei seinem schließlich maßgeblichen städtebaulichen Plan durch die Londoner Docklands inspirieren. Ähnlich rigoros wie Rotterdams Nachkriegsplanung verzichtete der Vorschlag, der zur Grundlage des Masterplans wurde, weitgehend auf die Erhaltung von historischer Bauten. Bezeichnenderweise blieben alle abweichend dazu vorgesehenen Umnutzungen historischer Hafenbauten (Architekten: Benthem & Crouwel, Fumi Hoshino) bisher unrealisiert.

that has been applied with full force during my lifetime but in fact had begun earlier, in the second half of the 19th century."
In the first days of the war the historic core of the city was almost completely destroyed except for the town hall, main post office, stock exchange and the Laurenskerk without an air raid alarm having been sounded – a tragic chain of negligence and misunderstandings, following the German ultimatum. After the end of the war all sites in the ruined area were compulsorily purchased, the streets and all the infrastructure removed, even underground, creating a tabula rasa for drawing board planning whose synthetic urban design was proclaimed a failure only 30 years later.
Against such a background, the goal of the city on the Maas peninsula of Kop van Zuid had to be the realisation of a high functional level in the form of innovative architecture. After Rotterdam's having been regarded for decades as the 'new town' par excellence, its extension is introduced, completely without embarrassment, as the "City of Tomorrow". But apart from a few buildings by Kees Christiaanse, Erick van Egeraat, Cepezed BV, Kruisheer/Elffers with Peter Wilson and the extremely coherent collaboration between Bolles + Wilson that extends across a series of smaller and larger project, the results are sobering in the extreme. The expansive gesture of the Erasmus Bridge has not been taken up and there is not much time left until completion in 2010. On a prime (also cleared) urban site, additionally surrounded and intersected by water, urban planning has remained schematic, non-spatial and unapproachable, and the architecture has in most cases remained uninspired. This is all the more surprising given the bad experiences in the former port area of the neighbouring western peninsula Katendrecht, where the port basin was filled in in the mid 1980s, according to the tabula rasa principle, to create a monofunctional residential area for which the river is only an urban edge or backyard.
After all, enthusiam for a city on the water still exists among the public and can be proven by numerous examples. It only just proved possible to prevent the canals on Kop van Zuid from being filled in. On 27 August 1991 a system of quality control with wide-ranging powers at all levels of the planning process was established under the leadership of the landscape architect and town planner Riek Bakker (Bakker & Bleeker), who has already received the Maaskant prize for her commitment and been director of the city planning office since the decisive year 1986. As well as Dutch experts this quality team includes a Spaniard and a Swede and also Mike Guyer from Gigon/Guyer.
With the move of the port activities to Europort, the remaining residential areas inhabited by the port workers declined as a result of high unemployment and criminality into an area in need of complete urban renewal. In concrete terms the planning of von Kop van Zuid began with an architecture symposium in 1982, interestingly organised by the Rotterdam Art Foundation, in order to expose the architectural monotony of the first buildings. The architects involved: Aldo Rossi, Josef Paul Kleihues, Oswald Mathias Ungers and Derek Walter were set for the first time the task of planning the entire old port area on the south bank of the Maas in 1987. Teun Koolhaas, Rem Koolhaas' nephew, drew his inspiration for his urban plan, which was ultimately to be decisive, from the London docklands. With a rigour similar to post-war planning in Rotterdam, the suggestion that formed the basis for the master plan dispensed largely with the retention of existing buildings. Characteristically, all suggestions that differed in some way and suggested the reuse of historical port buildings (architects: Benthem & Crouwel, Fumi Hoshino) have remained unrealised to date.

Kees Christiaanse:
Wohnblock | housing block, **Stadstuinen**

Erick van Egeraat:
Ichthus College

Fumi Hoshino:
„Pakhuismeesteren", Wilhelmina-Pier, Umnutzung und Aufstockung eines historischen Hafenspeichers
Conversion of and additional storeys for a historic harbour warehouse

Autopoesie

In der Realisierung setzten die Stadtplaner Rotterdams anders als Meinhard von Gerkan in China nicht auf eine restriktive Gesamtplanung, sondern auf die Heterogenität überzeugender und weniger überzeugender Einzelplanungen. Charakteristikum von Kop van Zuid ist darum nicht die Architektur, sondern die Struktur der langen Binnenhöfe senkrecht zum Fluss, die in der Planung von Teun Koolhaas anders als im städtischen Masterplan weitgehend getilgt waren. Die Distanz zwischen massiven Bürotürmen und Wohnburgen auf der einen und kleinformatigen Hafenelementen wie Schleusen, Brücken und Kränen auf der anderen Seite behindert jedoch eine denkbare urbane Atmosphäre dieses Standorts, wie sie konkurrierenden Planungen Amsterdams sogar als reiner Wohnstandort gelungen ist. Nicht zum ersten Mal zeigt sich die Urbanität Rotterdams heterogen und chaotisch. Städtebau und Architektur entwickeln einen unversöhnlichen Gegensatz. Die glänzenden neuen Türme voller Selbstvertrauen erheben sich über eine Stadt, in der das Scheitern des Projekts Moderne unverkennbar bleibt, ohne diesen Anspruch auch nur architektonisch einzulösen.
Eines der wichtigsten Elemente, das mit der labyrinthischen Stadt verlorenging, ist die Kategorie des Bildes. Die visuelle Reorganisation von Orten ist auf Kop van Zuid weitgehend gescheitert. Damit aber bleibt auch die Identität des Stadtteils verschwommen. Riek Bakkers städtebauliches Konzept der „langen Linien" und Sichtachsen in zentralen Stadtteilen, gerahmt von großmaßstäblichen Hochbauten, wie auch das Konzept der „Architektenfamilien" hat diesem Mangel wenig entgegenzusetzen. Ein rettender Mechanismus sich selbst generierender städtischer Poesie, wie er Manhattan gelingt, hat sich an der Maas noch nicht eingestellt.

Automatic Poetry

In the realisation phase the town planners in Rotterdam have, in contrast to Meinhard von Gerkan in China, not used a restrictive overall plan but have permitted the heterogeneity of convincing and less than convincing individual plans. The characteristic feature of Kop van Zuid is therefore not the architecture but the structure of the long internal courtyards at right angles to the river, which in the plan by Teun Koolhaas, in contrast to the urban master plan, were largely eliminated. The distance between massive office towers and apartment blocks on the one hand, and the small scale port elements such as locks, bridges and cranes on the other, prevents any conceivable urban atmosphere developing such for an entirely residential area – as succeeded in rival plans in Amsterdam. Not for the first time the urbanity of Rotterdam is both heterogeneous and chaotic. Urban planning and architecture develop an irreconcilable opposition. The shiny new towers rise full of self-confidence above a city in which the failure of the project of modernism remains irrefutable without ever having given this insight an architectural form.
One of the most important elements that was lost with the labyrinthine city is the category of the image. The visual reorganisation of places has, for the most part, failed on Kop van Zuid. Therefore the identity of this urban district is blurred. Riek Bakker's urban concept of "long lines" and visual axes in central city districts, framed by large scale high-rise buildings as also the concept of "architect families" can do little to counter these deficits. A saving mechanism based on a self-generating urban poetry, such as succeeded in Manhattan, has not yet occurred on the Maas.

Der 1999 erschienene Band „Kop van Zuid 2" von Piotr Ostojski Ostoja und Willem van der Laan schildert den planerischen Ansatz, ist jedoch in seinen Projektdarstellungen veraltet. Diese im Buchhandel vergriffene Publikation wie alle aktuellen Informationen sind im Kop van Zuid Information Centre (www.kopvanzuid.rotterdam.nl) zu bekommen.

The book *Kop Van Zuid 2* by Piotr Ostojski Ostoja and Willem van der Laan, published in 1999, describes the starting points for the planning but its illustrations of projects are out of date. This out-of-print publication as well as all current information is available from the Kop van Zuid Information Centre (www.kopvanzuid.rotterdam.nl).

1 Stichting Wolkenkrabbers Rotterdam / Rotterdam Skyscrapers Foundation (Hrsg.): High-rise is a Must! (Edition Magazine The slender City 4), Rotterdam 2000
2 Martin Aarts, Ben Maandag, Jos Stoopman u.a.: Accelerating Rotterdam / Stad in Versnelling, Rotterdam 2000
3 DeutschlandRadio Berlin 24.März 2002
4 Vgl. Hilde de Haan, Ids Haagsma: De architectuur van Erasmus, een karakterschets van de Nederlandse Bouwkunst, Haarlem 1990
5 Christian Welzbacher: Polderland unter. Die Niederlande vor der Wahl: Konservatismus ist in - auch in der Architektur, DIE ZEIT 20/2002
6 Hans Ibelings: Supermodernism. Architecture in the Age of Globalization, Rotterdam 1998/2003
7 de Architekten Cie. (Hrsg.): Profession Architect. de Architekten Cie., Rotterdam 2002

1 Stichting Wolkenkrabbers Rotterdam/Rotterdam Skyscrapers Foundation (Ed.): 'High-Rise Is A Must!' (*The Slender City* magazine edition No. 4), Rotterdam 2000
2 Martin Aarts, Ben Maandag, Jos Stoopman and others: *Accelerating Rotterdam/Stad in Versnelling*. Rotterdam 2000
3 DeutschlandRadio, Berlin 24 March 2002
4 Cf. Hilde de Haan, Ids Haagsma: *De Architectuur Van Erasmus, Een Karakterschets Van De Nederlandse Bouwkunst*. Haarlem 1990
5 Christian Welzbacher: 'Polderland Unter. Die Niederlande Vor Der Wahl: Konservatismus Ist In – Auch In Der Architektur', *Die Zeit* 20/2002
6 Hans Ibelings: *Supermodernism. Architecture In The Age Of Globalization*. Rotterdam 1998/2003
7 de Architekten Cie. (Ed.): *Profession Architect*. de Architekten Cie., Rotterdam 2002

Herzog & de Meuron

Herzog & de Meuron

Geschäftsviertel „Fünf Höfe" in München, Deutschland
Fünf Höfe Commercial District in Munich, Germany

Photos Margherita Spiluttini, Joachim Hofer
Text Wolfgang Jean Stock

Städtebau nach innen

Wer vom Isarhochufer aus die Münchner Stadtsilhouette in den Blick nimmt, sieht zwischen den vertrauten Türmen und Giebeln einen Wald von Baukränen. An vielen Stellen wird die bayerische Landeshauptstadt umgestaltet – und mitten im Zentrum liegen die „Fünf Höfe", deren größerer Südteil als erster fertig gestellt wurde. Die Pritzker-Preisträger Jacques Herzog und Pierre de Meuron verstehen ihre städtebauliche Intervention als europäische Alternative zum Typus der amerikanischen Shopping-Mall.

Urban Planning Directed Inwards

If you look at the urban silhouette of Munich from the high bank of the Isar you see a forest of building cranes between the familiar towers and gables. The Bavarian capital is undergoing a redesign at a number of points. At the centre lie the Fünf Höfe (Five Courts), of which the larger, southern part was the first to be completed. Pritzker prize-winners Jacques Herzog and Pierre de Meuron see their urban intervention as a European typological alternative to the American shopping mall.

FÜNF HÖFE
PERUSAHOF

Alte Fassaden (o.)
neu formulierte Prannerpassage (re.)
Old facade (top)
Reformulated Prannerpassage (right)

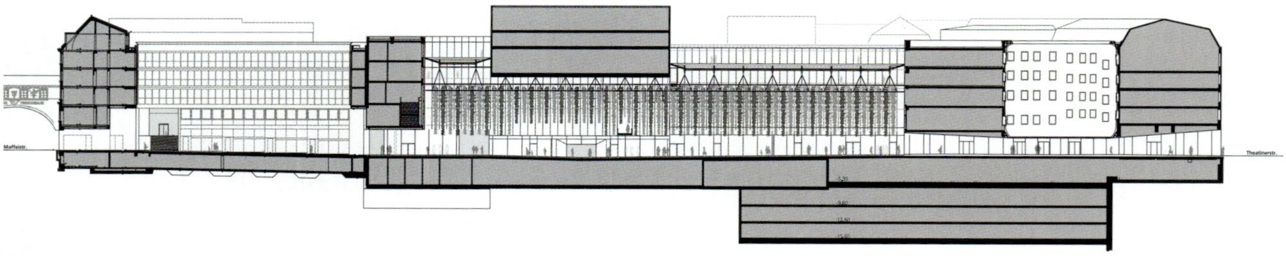

Längsschnitt durch Maffeihof und Salvatorpassage
Longitudinal section through Maffeihof and Salvatorpassage

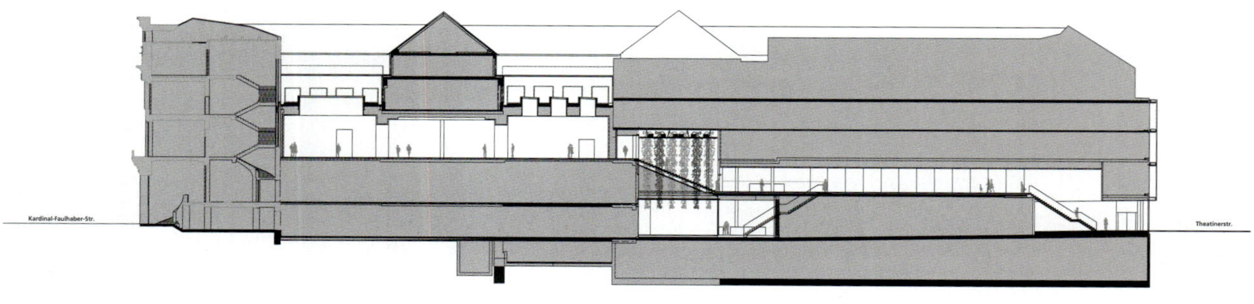

Querschnitt
Section

SALVATORPASSAGE

Ursprünglich hieß es: Tabula rasa. Nahezu ein ganzes Quartier im traditionellen Münchner Bankenviertel wäre umgepflügt worden, hätte sich der radikale Entwurf durchgesetzt, mit dem Jacques Herzog und Pierre de Meuron im Frühjahr 1994 den internationalen Wettbewerb zur Umgestaltung des verschachtelten Hypo-Blocks gewonnen hatten. Weshalb diese Planung zu Fall kam, ist bis heute nicht eindeutig geklärt.
Dass die Münchner nach wie vor – und ganz besonders im Bereich der Altstadt – moderner Architektur gegenüber reserviert eingestellt sind, weil sie an vertrauten Bildern hängen, dafür lieferte das Tauziehen um die Erhaltung der Hypo-Fassaden an der Theatinerstraße ein geradezu klassisches Beispiel. Da bedurfte es gar nicht mehr irgendwelcher Auflagen der Denkmalpflege – den Protest regelte die bürgerschaftlich aktive Öffentlichkeit von selber. Als nach der Fusion von Hypo- und Vereinsbank im Jahr 1997 das Projekt wieder aufgenommen wurde, hatten sich aber auch die Vorgaben des Bauherrn geändert. Durch die Fusion wurde der Block auf einer Fläche von 20.000 Quadratmetern nicht mehr für eigene Zwecke benötigt. Deshalb wurden Herzog & de Meuron mit einem neuen Entwurf beauftragt, der beides garantieren sollte: eine optimale Verwertung des Grundstücks durch Läden, Gastronomie und Büroflächen wie auch die Zustimmung der kritischen Bürgerschaft.

Originally the idea was to make a *tabula rasa*. Had the radical design, with which Jacques Herzog and Pierre de Meuron won the international competition for the redesign of the cramped Hypo block in 1994, been carried out this would have meant ploughing up almost an entire quarter of Munich's traditional banking district. The reason this plan was not executed remains unclear to the present day.
The tug of war that took place over the preservation of the Hypo facades along Theatinerstrasse is a classic illustration of the fact that the citizens of Munich have a reserved approach to modern architecture –particularly in the old town as they are attached to its familiar images. Constraints imposed by the conservation authorities were not necessary as the public organised the protest itself. When the project was taken up once again following the merger between the Hypobank and the Vereinsbank in 1997 the client's requirements had also altered. As a result of the merger, a block with a surface area of 20,000 square metres was not required for the bank's own use. As a result Herzog and de Meuron were commissioned to deliver a new design which was to guarantee optimum use of the site by shops, restaurants and offices and which was also to secure the approval of the critically inclined citizens.

Durchdringung von Höfen und Passagen
Interweaving of courtyards and passages

SAN PELLEGRINO
SAN PELLEGRINO

100 Jahre – Drei Häuser

Anfang 1998 stand das Ergebnis unter dem neuen Namen „Fünf Höfe" fest: Anstelle der ursprünglichen Kahlschlag-Sanierung durch große Neubauten, die Herzog & de Meuron etwas martialisch als „schwere Steine im Quartier" angekündigt hatten, werden die alten Häuser entkernt. Durch die ins Blockinnere gerichteten städtebaulichen Eingriffe bleiben 60 Prozent des Bestands erhalten, weitgehend auch die umkämpften Fassaden. Unstrittig war jedoch das von den Architekten im Wettbewerb vorgeschlagene Prinzip, die Gebäude durch ein dichtes Netz aus Passagen und Höfen miteinander zu verbinden. Damit findet das seit einem Jahrhundert währende Um- und Weiterbauen an diesem Block einen glücklichen Abschluss. Im Abstand von jeweils 50 Jahren wird dann das ganze Quartier dreimal grundlegend verändert worden sein.

1898 hatte die Bayerische Hypotheken- und Wechsel-Bank dort ihren neobarock gestalteten Stammsitz bezogen. Nachdem er im Zweiten Weltkrieg zu 95 Prozent zerstört worden war, ließ die Bank nach Plänen des Architekten Adolf Abel an der Theatinerstraße bis 1953 ein neues Gebäude errichten, dessen lang gestreckte Fassade eine konservative Interpretation des Purismus der 1920er Jahre darstellte. Bedeutsam war dabei ein struktureller Eingriff: die Hypo-Passage. Mitsamt ihrem Hof bildete sie einen ersten Nachhall auf Abels zukunftsweisenden Aufbauplan für die innere Stadt aus dem Jahr 1947. Dessen Ziel hatte darin bestanden, die Baublöcke durch eine abwechslungsreiche Folge von Plätzen, Höfen und Passagen zu öffnen. Abels Planung war jedoch mit dem in München nach 1945 unveränderten Bodenrecht unvereinbar und verschwand deshalb in den Schubladen.

100 Years – Three Buildings

At the beginning of 1998 the result was decided upon, it bore the new name: Fünf Höfe. Instead of the originally planned total redevelopment using large new buildings which Herzog and de Meuron (in a somewhat martial manner) had announced as "weighty stones in the quarter", the old buildings were hollowed out. The fact that the urban interventions were directed towards the interior of the block meant that 60% of the original substance was preserved as were, for the most part, the hotly defended facades. It was agreed, however, that the principle of connecting the buildings by a dense network of arcades and courtyards suggested by the architects in their competition entry should be retained. So the development of and expansion to this block, which has continued for a century, has reached a happy conclusion. At intervals of fifty years the whole quarter has been fundamentally altered three times.

In 1898 The Bayerische Hypotheken- Und Wechsel-Bank moved into its neo-baroque headquarters on this site. After 95% was destroyed in the Second World War the bank erected a new building along Theatinerstrasse to a design by architect Adolf Abel completed in 1953. Its long facade represented a conservative interpretation of 1920s Puritanism. One structural intervention was of particular significance: The Hypo Arcade. Together with its courtyard it was a first echo of Abel's forward-looking development plan for the inner city dating from 1945. His goal was to open up the blocks by means of a varied sequence of squares, courtyards and arcades. However Abel's plan proved impossible according to the ground law in Munich which remained unaltered after 1945 and so it ended up in the plan chest.

Fassaden als Lichtfilter im Aluminium-„Kettenhemd"
Facade as light filter in the aluminium 'chain-mail shirt'

Spiegel- und Schichtungseffekte
Reflective and layered effect

SALVATORPASSAGE
CAFÉ KUNSTHALLE

Perusapassage mit Luftraum
Perusapassage with open-air space

Bauabschnitte, Wegführungen und Neubauten (o.), Blockstruktur, Grundstücksverteilung und Abbruch (u.)
Building phases, access routes and new buildings (above), Block structure, properties and demolishing (bottom)

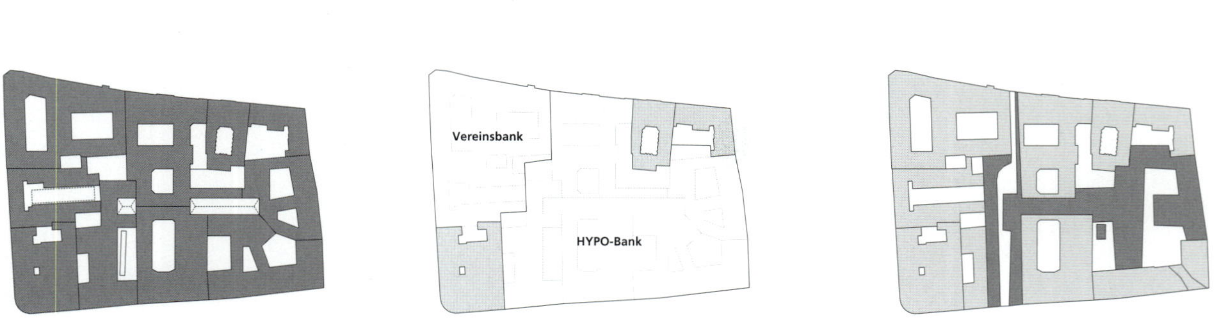

SCHNEE
BACH
TAL
WALD
GEBIRGE
GLETSCHER
FELSEN
WALD

Raumfolgen, Lochblenden, Light Boxes

Nun, da der Großteil der „Fünf Höfe" fertig gestellt ist, lässt sich erkennen, dass Herzog & de Meuron die Vision von Abel weit übertroffen haben – trotz struktureller Nähe sind Raumbildung und Formensprache das glatte Gegenteil von Abels bodenständigen Veduten. Herzog & de Meuron haben bei ihrem „Städtebau nach innen" auf die sinnliche Wirkung der Räume gesetzt, auf den Wechsel von Enge und Weite bei der Wegführung wie auf die Folge von gedeckten und zum Himmel hin offenen Höfen, so dass die Außenwelt mit Sonnenschein, Regen oder Schnee immer wieder präsent wird.

Nur an einer Stelle bildet sich die neue Nutzung nach außen ab: Auf Wunsch des Bauherrn wurde zwischen die historischen Fronten an der Theatinerstraße eine neue Fassade eingeschnitten, deren unverhohlen zeitgenössische Gestalt die Blicke auf sich zieht. Diese mehrschichtige Hauswand mit Faltläden aus Lochblech, die sich in geschlossenem Zustand wie ein glitzerndes Gewebe vor die verglasten Geschoße legen, prägt als Thema auch die Innenhöfe. Die flexibel ausfahrbaren Läden sorgen aber nicht nur für lebendige Wände, sondern dienen zugleich als Sonnen- und Sichtschutz. Besonders reizvolle Effekte erzeugt die Gebäudehülle – von den Münchnern schon als „Kettenhemd" tituliert – beim Blick von drinnen nach draußen, weil man wie durch einen feinen Filter in die Umgebung schaut.

Mit der neuen zeichenhaften Fassade hat die behäbige Theatinerstraße eine wohltuende Spannung erhalten. Zugleich weist sie auf den Eingang zur beliebten Hypo-Kunsthalle hin, die ins zweite Obergeschoß verlegt wurde, damit die unterschiedlich großen Säle möglichst viel natürliches Oberlicht erhalten. Alle acht Schauräume sind mit ausgewogenen Proportionen, schlichten Durchgängen, klar definierten Wänden und einer verborgenen Haustechnik so beschaffen, wie es Künstler und Kuratoren stets gefordert haben. Ein wesentliches Element sind die von Herzog & de Meuron für die Londoner Tate Modern entwickelten und jetzt verfeinerten „Light boxes": Durch transluzente, nahtlose Folien fällt das Tages-, Kunst- oder Mischlicht gleichmäßig und fast schattenfrei über die Wände.

Spatial Sequences, Perforated Screens, Light Boxes

Now that the larger southern part of the Fünf Höfe has been completed it can be appreciated that Herzog and de Meuron have surpassed Abel's vision. Despite the structural closeness, the formation of space and the formal language they employ are the direct opposite of Abel's down-to-earth vedutas. Herzog & de Meuron's inward-oriented urban planning places emphasis on the sensual effect of space, on the alternation between narrow and broad, on a system of routes and on a sequence of covered and open courtyards so that the outside world is always present whether in the form of sunshine, rain or snow.

The new function is revealed externally at only one point. In response to the client's wish, a new facade was incised between the historic fronts along Theatinerstrasse, the unashamedly contemporary design of which is eye-catching. This multi-layered wall with folding shutters of perforated metal which, when closed, are like a glittering mesh in front of the glazed storeys is a theme that also dominates the inner courtyards. The shutters, which can be moved outwards flexibly, not only provide animated walls but also offer visual and solar screening. The building shell, already dubbed 'the chain-mail shirt' by the citizens of Munich, provides a particularly enchanting effect when one looks outwards from inside: It is like peering at the surroundings through a fine filter.

The new symbolic facade has lent the rather staid Theatinerstrasse a certain tension which does it good. It also indicates the entrance to the popular Hypo Kunsthalle gallery, which was moved to second floor level to allow the halls of various sizes as much natural top lighting as possible. All eight gallery spaces have well balanced proportions, simple connecting passages, clearly defined walls and concealed services made in the way constantly demanded by artists and curators. The 'Light Boxes' developed by Herzog & de Meuron for the London Tate Modern and refined further here are an important element. Daylight, artificial light or combined light falls along the walls through translucent, seamless foil evenly and almost without shadows.

Ausstellungssaal in der Kunsthalle
Exhibition hall in the arts centre

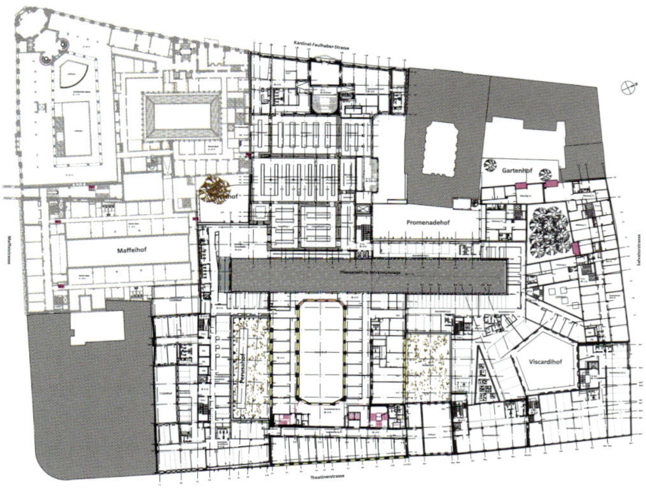

2. Obergeschoß
2nd level

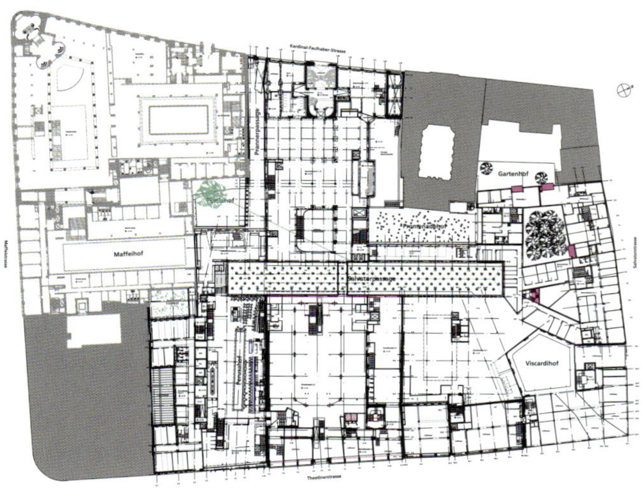

1. Obergeschoß
1st level

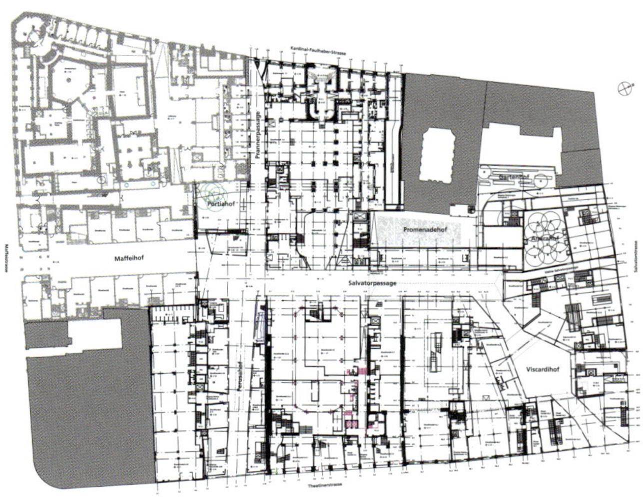

Erdgeschoß
Ground floor

Salvatorpassage

Sinnliche Effekte

Die Lounge vor der Kunsthalle öffnet sich zum Geviert der Höfe und Passagen, die sich in ihrer Gestalt bewusst unterscheiden. Von der Theatinerstraße aus erreicht man zunächst den teilweise überdeckten Perusahof, wo Spiegelungen auf den grünlichen Glaspaneelen die Konturen zwischen oben und unten wie auch zwischen innen und außen verwischen. Keineswegs immateriell wirkt dagegen die anschließende Salvatorpassage, die zur Hälfte fertig gestellt ist. Als sinnliches Gegenstück zum sachlichen Ambiente, als „Natur in der Architektur" ranken dort tropische Pflanzen aus Trögen über dem hohen Gitterrost in den Raum hinunter. Die Folge der Erlebnisräume mündet schließlich in der engen Prannerpassage, die von Herzog & de Meuron überraschend poetisch gestaltet wurde: Hier funkeln putzbündig eingelassene Glaspailletten im Licht. Gemeinsam ist allen Bereichen die Transparenz – durch die voll verglasten Wände geht der Blick hinein in die Geschäfte und gastronomischen Betriebe.
Für München sind die „Fünf Höfe" ein Zugewinn an ebenso eleganter wie intelligenter Architektur. Ihre Bedeutung reicht aber weiter, denn sie haben einen Modellcharakter für europäische Stadtsanierung haben. Die neue Nutzung in der Mischung von Handel und Kultur, Gastronomie und Wohnen sorgt dafür, dass dieses zentrale Quartier auch nachts nicht verödet. So sind die behutsam in den Bestand integrierten „Fünf Höfe" eine überzeugende europäische Antwort auf die amerikanischen Shopping-Malls. Womit sich denn auch die heftige Einmischung der Bürgerschaft im Nachhinein als positiv erwiesen hat, wie Jacques Herzog freimütig betont: „Wir trauern der alten Lösung nicht nach."

Sensual Effects

The lounge in front of the Kunsthalle opens onto the mesh of the courtyards and arcades which deliberately differ considerably in design. From Theatinerstrasse one first reaches the partly covered Perusahof where the reflections of the greenish glass panels blur the contours between above and below and between inside and outside. In contrast the adjoining Salvatorpassage, half of which has been completed, is anything but immaterial. As a sensuous contrast to the cool ambiance, as a piece of "nature amidst the architecture" tropical plants creep into the space below from troughs along a tall trellis. The sequence of spaces leads finally into the narrow Prannerpassage, which was surprisingly designed in a most poetic fashion by Herzog & de Meuron. Glass pieces set flush with plaster sparkle in the light. Transparency is a quality common to all the different areas, one can look into the shops and restaurants through walls glazed the full height.
For the people of Munich the Fünf Höfe represent a gain in terms of elegant and intelligent architecture. But their significance goes beyond this: They are a model for the rehabilitation of European cities. The new use and the mix of business and culture, restaurants and residential means that this quarter does not become a desert at night. The Fünf Höfe are carefully integrated in the existing substance and offer a convincing European response to the American Shopping Mall. In this case the hefty interference by the citizens has, in retrospect, turned out to be a positive thing. As Jacques Herzog freely emphasises: "We don't mourn the earlier solution".

Technische Daten | technical data

Fünf Höfe
München/Deutschland, Theatiner-/Salvator-/Kard.-Faulhaber-Straße

Bauherr | client **Bayerische Hypotheken- & Vereinsbank AG, München**
Planung | planning **Herzog & de Meuron**
Projektteam | project team **Jaques Herzog, Pierre de Meuron; Robert Hösl, Tim Hupe; Sascha Arnold, Andrea Bernhard, Jean-Claude Cadalbert, Eric Frisch, Martin Fröhlich, Susanne Kleinlein, Susanna Knopp, Katharina Kovarbasic, Ursula Müller, Fabian Ochs, Heiner Reimers, Peter Reinhardt, Florian Schmidhuber, Markus Wassmer; Konstanze Beelitz, Rolf Berninger, Silvia Beyer, Enrica Ferrucci, Agnes Förster, Milena Kondoferska, Jan Kurz, Julia Lingenfelder, Jan Frederik Peters, Matthias Pektor, Daniel Reisch, Christoph Röttinger, Christian Schühle, Korinna Thielen**
Kunst am Bau | artwork **Rémy Zaugg, Basel; Thomas Ruff, Düsseldorf**
Statik | structural consultant **Obermeyer Planen + Beraten, München**
Fassadenplanung | facade planning **Memmert & Partner, Neuss**
Akustikplanung | acoustic concept **Möhler & Partner, München**
Lichtplanung | lighting concept **Peter Andres (Kunsthalle, Passagen)**
Landschaftsplanung | landscape architects **Burger Landschaftsarchitekten, München**

Nutzfläche | floor area	78.000 m²
Bebaute Fläche | built-up area	13.500 m²
Umbauter Raum | cubage	340.000 m³
Wettbewerb | competition	1994
Planungsbeginn | start of planning	1995
Baubeginn | start of construction	1999
Fertigstellung 1. Bauabschnitt | completion, phase 1	2/2001
Fertigstellung Kunsthalle | completion, arts centre	6/2001
Gesamteröffnung | opening	2003
Baukosten | building costs	ca. 138 Mio EUR

Lab architecture studio

Lab architecture studio

Scherben-Kunst – Melbournes Federation Square als urbane Naht
Splinter Art – Melbourne's Federation Square: The Repair Of An Urban Fracture

Photos John Gollings
Text Viviane Stappmanns

Bis vor kurzem wuchs in Melbourne, angetrieben durch stetige Zuwanderung, einer der größten Vorstadtteppiche der Welt heran. Während die Stadt nach außen wucherte, ließen die Pendler den Central Business District abends als ausgestorbenes Zentrum zurück. Das nun überplattete Gebiet des Federation Square teilte zudem die Stadt durch eine große Brache in zwei Hälften. Der neue Platz erfüllt mehrere wichtige städtebauliche Aufgaben: Er wurde nicht nur als Stadtteil-verbindender Freiraum geplant, sondern generiert durch Ansiedlung mehrerer Kulturinstitutionen auch neues urbanes Leben. Heute befinden sich dort die Nationalgalerie des Staates Victoria, das Australian Centre for the Moving Image, eine Produktionsstätte für den staatlichen Fernsehsender SBS, mehrere Restaurants und Pubs, sowie Geschäfte und eine multifunktionale „Plaza".

Until recently one of the world's largest carpets of suburban sprawl grew steadily around Melbourne due to the steady stream of immigrants. While the city expanded outwards, in the evening the exodus of commuters from the Central Business District left the city centre dead. In addition, a large railway area under Federation Square separated the city into two halves. The new square fulfils several important tasks in terms of urban planning: it was not envisaged just as an open space linking two urban districts but, as a result of the decision to locate a number of cultural institutions there, it generates new urban life. Today the National Gallery of the State of Victoria is located there along with the Australian Centre for the Moving Image, a production centre for the state television station SBS, several restaurants and pubs as well as shops and a multifunctional 'plaza'.

SBS

Vom Umgang mit Architekten

Die Taxifahrer in Melbournes Innenstadt haben sie, genauso wie die Restaurantangestellten, Ladenbesitzer, Straßenbahnfahrer und Gemüseverkäufer: eine Meinung über den Federation Square, jenen Teil der Innenstadt, der das Schachbrett-Muster der Straßen und Gassen Melbournes um ein Feld erweitert. Melbournians hassen oder lieben den Federation Square. Die Gebäudeanordnung zu einem öffentlichen Raum, der sich über insgesamt 3,8 Hektar zieht und das Zentrum mit dem die Stadt teilenden Fluss Yarra verbindet, fordert klare Aussagen.

Seit Oktober 2002 ist der Federation Square offiziell ein Teil der Vier-Millionen-Einwohner-Metropole. Das Londoner Architekturbüro Lab architecture studio, bestehend aus dem Texaner Donald Bates und dem Australier Peter Davidson, gewann 1997 den ausgeschriebenen Wettbewerb für die Bebauung einer zu konstruierenden Plattform über zwölf Bahngleisen der naheliegenden Bahnstation. In der zweiten Phase der Planung schlossen sich Lab mit Bates Smart, einer der ältesten australischen Architekturfirmen, zusammen und verlegten ihren Sitz nach Melbourne.

Die folgenden fünf Jahre des Baus waren angereichert mit Kontroversen über Budgets, Design, Änderungen des Designs, Meinungsverschiedenheiten zwischen Stadtverwaltung, Landesregierung, National Trust, den lokalen Medien und den Architekten.

Australiens Umgang mit Architekten, die Wettbewerbe gewinnen und dann ihre Projekte nach längerem Tauziehen mit kurzsichtigen Bürokraten unfertig verlassen, ist legendär. Jørn Utzon, Designer des Opernhauses, verließ das Land 1966 und schwor, niemals zurückzukommen. Nach einer öffentlichen Entschuldigung der Regierung und einer Einladung, die Innenarchitektur des Gebäudes neu zu gestalten, beaufsichtigt nun der Sohn des Architekten vor Ort die Umsetzung der Pläne seines Vaters, der inzwischen auf Mallorca lebt. Auch Walter Burley Griffin, Planer und Architekt der Hauptstadt Canberra, legte seine Arbeit frühzeitig nieder. Bates und Davidson jedoch harrten aus.

Neuer Stadt-Raum

Die internationale Ausschreibung des Projekts forderte den Entwurf eines öffentlichen Raums, der bis zu 15.000 Besucher in einem Open-Air-Amphitheater aufnehmen konnte. Ein weiterer Aspekt reizte die Architekten besonders: „Die Ausschreibung rief zum Design eines neuartigen öffentlichen Raums für das 21. Jahrhundert auf, in dem kulturelle und kommerzielle Einrichtungen zusammen wirken. Diese außergewöhnliche Aufgabe war einer der Gründe, warum wir es gemacht haben," sagt Davidson in seinem Büro, 100 Meter Luftlinie vom „Fed Square" entfernt.

Heute befinden sich dort die Nationalgalerie des Staates Victoria, die dem Projekt erst nach dem Wettbewerb hinzugefügt wurde, das Australian Centre for the Moving Image (ACMI, eine staatliche kulturelle Organisation zur Förderung von Film, Fernsehen und neuen Medien), eine Produktionsstätte für den staatlichen Fernsehsender SBS, mehrere Restaurants und Pubs, sowie Geschäfte und das „plaza", eine Versammlungsstätte für Anlässe wie Konzerte und Demonstrationen, oder zum informellen Essen, Trinken und Treffen.

Ins Leben gerufen wurde der Wettbewerb für Federation Square durch den ehemaligen Premier Victorias, Jeff Kennett, der es sich zum Ziel gesetzt hatte, Melbournes Stadtbild attraktiver zu machen. Mit der Eröffnung des Square sollten die Feierlichkeiten zum 100jährigen Jubiläum der australischen Föderation begangen werden. Der Name „Federation Square" ist eigentlich missverständlich. Nichts an dem Pro-

On Dealing With Architects

The taxi drivers in Melbourne's inner city have one – as do restaurant staff, shop owners, tram drivers and people selling vegetables: an opinion on Federation Square, the part of the inner city that extends the chequerboard pattern of streets and lanes in Melbourne by a further square. Melbournians either hate or love Federation Square. The positioning of buildings around an open space that extends over a total of 3.8 hectares and connects the city centre with the river Yarra, which divides the city, demands unambiguous statements.

Since October 2002 Federation Square has officially been part of the four million people metropolis. In 1997 the London architects Lab architecture studio, made up of Texan Donald Bates and Australian Peter Davidson, won the competition set up for the development of a platform to be built above 12 of the railway tracks of a nearby railway station. In the second phase Lab joined forces with Bates Smart, one of Australia's oldest architecture practices, and moved their office to Melbourne.

The following five years of construction were marked by controversies about budgets, design, changes to the design, and differences of opinion between the city administration, the state government, the National Trust, the local media and the architects.

Australia's treatment of architects who win competitions and then abandon their projects after years of struggle with shortsighted bureaucrats is notorious. Jørn Utzon, designer of the Sydney Opera House, left the country in 1966 swearing never to return. After an official apology by the government and an invitation to redesign the interior of the building the architect's son is now supervising on site the realisation of plans by his father, who now lives in Majorca. Walter Burley Griffin, planner and architect of the Australian capital Canberra, also gave up his work prematurely. Bates and Davidson, however, stayed on.

New Urban Space

The international competition required the design of a public space that could accommodate up to 15,000 visitors in an open-air amphitheatre. A further aspect particularly attracted the architects: "The brief was committed to a new kind of 21st century civic space that saw cultural facilities and commercial facilities working together. This extraordinary task was one of the reasons we did it." Says Davidson in his office, 100 metres as the crow flies from "Fed Square".

Today this is the location of the National Gallery of the State of Victoria (which was first added to the brief after the competition), the Australian Centre for the Moving Image ACMI (a state funded cultural organisation to promote film, television and new media), a production centre for the state television corporation SBS, several restaurants and pubs as well as shops and the new 'plaza' – a place for events such as concerts and demonstrations or for informal dining, drinking and meeting people.

The competition for Federation Square was called into life by the former Prime Minister of Victoria, Jeff Kennett, whose aim was to make Melbourne's urban image more attractive. The opening of the square was to be part of the celebrations of the 100th anniversary of the Australian Federation. The name 'Federation Square' is somewhat misleading as nothing in this project, neither in plan nor in three dimensions is square. And Federation Square has little in common with the traditional market or town square. "It is a surprisingly un-urban site", says Davidson, "it was not simply about making an embedded urban square, because the site is effectively open."

SBS-TV- und Radiosender; Eingang zur National Gallery of Victoria
SBS TV & radio studios; entrance to the National Gallery of Victoria

jekt, weder auf dem Plan noch in dreidimensionaler Ansicht, ist „square", also viereckig. Auch mit dem „Town Square", dem Markt- oder Stadtplatz, hat der Federation Square wenig gemeinsam. „Die Lage ist überraschend un-urban," sagt Davidson, „es ging nicht nur darum, einen eingebetteten öffentlichen Platz zu schaffen, denn das Grundstück ist effektiv (nach zwei Seiten) offen."

Verwirrend und faszinierend

Die Architekten sprechen von einer „Architektur der Inklusivität", sie hielten sich an keine klassische Ordnung der Stadtplanung, sondern daran, das Leben der Straßen Melbournes, mit seinem Labyrinth aus Arkaden und Gassen zwischen den großen Verkehrsstraßen, zu erweitern und einen neuen Aspekt hinzuzufügen. Es ging nicht darum, eine Megastruktur, sondern ein lebendiges Stück Stadt zu gestalten.

Die beiden Architekten trafen sich während ihrer Zeit als Lektoren der Londoner Architectural Association (AA) Anfang der 1980er Jahre und gründeten ihr Büro 1994. Der Federation Square verließ als erstes gemeinsames Projekt das Zeichenbrett. Sie nennen ihre Zusammenarbeit „ein Labor der architektonischen Spekulation". Im Projekt Federation Square untersuchten die Architekten die Möglichkeiten der Integration von zeitgenössischer Geometrie und industrialisierter Fabrikations- und Konstruktionstechniken um konstruierbare, aber dennoch einzigartige Bauten zu gestalten.

Oder, wie es Davidson in Bezug auf die Fassade der Bauten pragmatischer ausdrückt: „Wir waren daran interessiert, ein Fassadensystem zu entwickeln, das industriell hergestellt werden konnte, uns aber trotzdem eine Differenzierung der Fassade jedes einzelnen Gebäudes erlaubte." Er verschwindet geschäftig in seinem Büro und kommt kurz darauf mit einer Grafik wieder, auf der die Fassade „auseinandergefaltet" dargestellt ist. „Wenn man sich die Konfiguration ansieht, fällt auf, dass sich ihr Maßstab ständig ändert."

Er vergleicht die Fassade mit einer M.C. Escher-Zeichnung, deren Original er sogleich hinter ein paar weiteren Zeichnungen hervorzieht. Die Identifikation der Gänse und Fische in „Luft und Wasser" als das eine oder andere ist für den Betrachter gleichzeitig verwirrend und faszinierend: „Wir wollten ihm diese Eigenschaft verleihen". Eine Architektur des „Unterschieds und der Übereinstimmung".

Confusing And Fascinating

The architects speak of "an architecture of inclusivity", they did not follow any classical system of town planning but aimed at expanding the life of the streets of Melbourne, with its labyrinth of arcades and lanes between the large traffic arteries, and adding a new aspect to it. The issue was not to design a mega-structure but a piece of living city.

The two architects met each other during their time as lecturers at the London Architectural Association (AA) at the start of the 1980s and founded their office in 1994. Federation Square was their first joint project to leave the drawing board. They call their collaboration "a laboratory of architectural speculation". In the Federation Square project the architects examined the possibility of integrating contemporary geometry and industrialised fabrication and construction technology to create buildings that are both unique and capable of being built.

Or, as Davidson puts it more pragmatically in reference to the facade of the buildings: "We were interested in developing a facade system that could be industrially manufactured but would allow us a differentiation of the facade with every building involved". He disappears busily into his office and comes out again shortly with a drawing on which the facade is represented 'unwrapped', so to speak. "If you look at the configuration it is constantly changing in terms of its scale".

He compares the facade with an M.C. Escher drawing, pulling out a copy of the original from behind a few drawings. Identifying the geese and fish in 'air and water' as one or the other is for the viewer both confusing and fascinating. "We wanted it to have that kind of quality". An architecture of "difference and coherence".

Australian Centre of the Moving Image: Grundriss (u.); fraktales Fassaden-Design (re.)
Floor plan (below); fractal facade design (right)

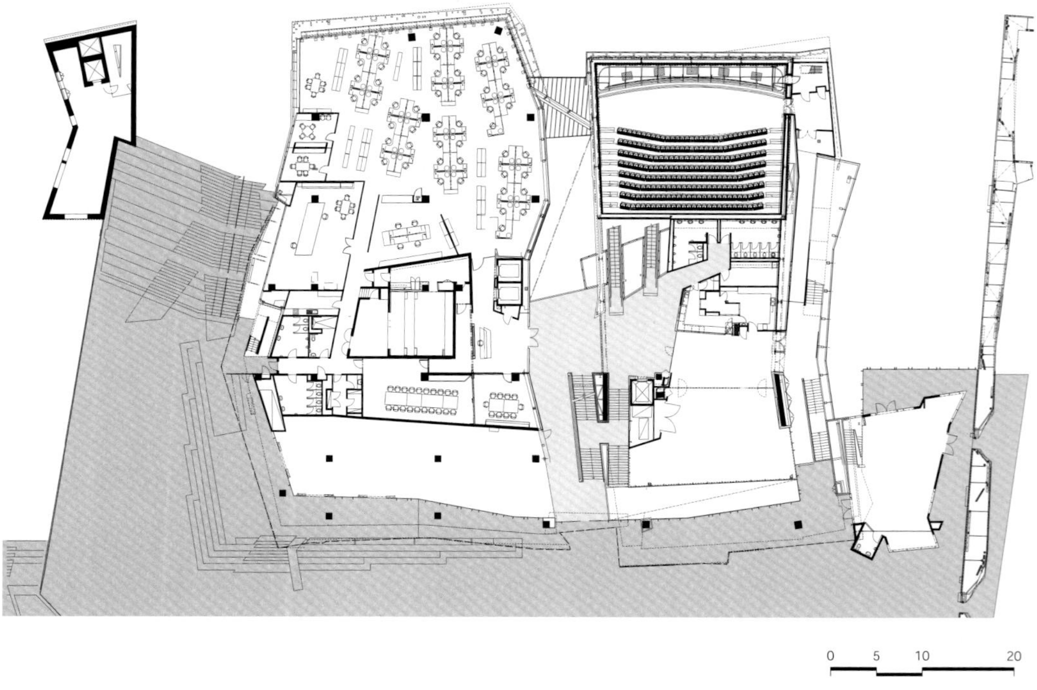

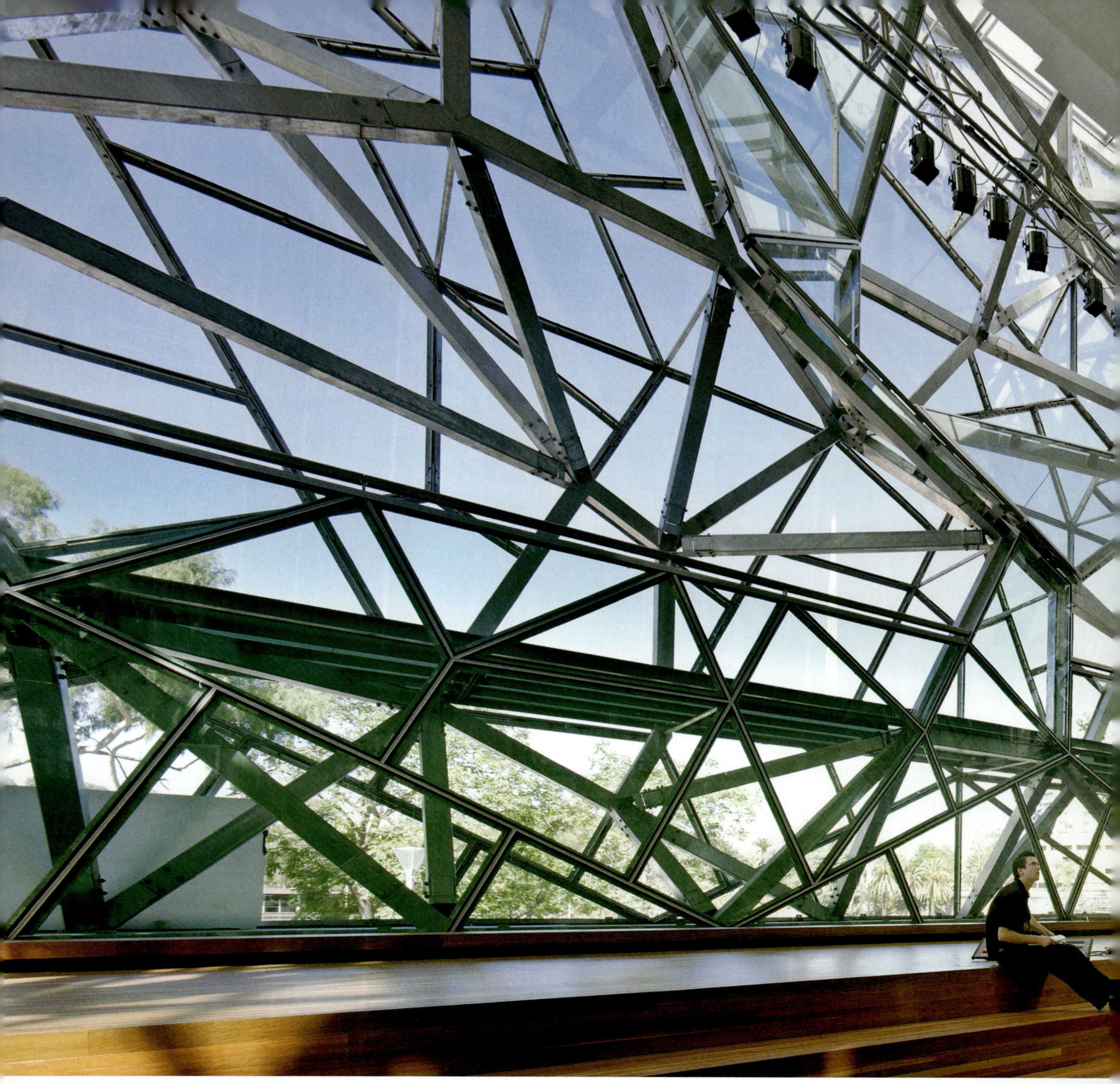

Die Scherbe als Zeichen

Drei verschiedene Materialien wechseln sich in den scheinbar fragmentierten Fassaden ab: Module aus Sandstein, Zink (perforiert und solide) und Glas, deren kleinstes eine Dreiecksform ist, setzen sich in variierenden Kombinationen zusammen. Fünf der Module formen in einem sogenannten „pinwheel-grid" größere Dreiecke, die „Panels". Fünf der „Panels" wiederum bilden die Konstruktionseinheit der „Megapanels".
Mit diesen Variationen traten die Architekten der Herausforderung entgegen, den programmatisch so unterschiedlichen Gebäuden Einheitlichkeit, aber dennoch Individualität zu verleihen. „Wir wollten die Oberflächen artikulieren, denn sonst wären die Gebäude einfach ausdruckslos gewesen," sagt Davidson.
Das Sandstein-Pflaster des „plaza" erinnert den Besucher an das große, unbekannte, rote Landesinnere. Die Architekten fanden den Stein in einem bereits geschlossenen Steinbruch in den Kimberley Ranges im Norden Australiens. Der Besitzer des Steinbruchs soll angeblich einmal mit Mustern seiner

The Splinters As A Symbol

Three different materials alternate in the apparently fragmented facades: modules of sandstone, zinc (perforated and solid) and glass, the smallest of which is a triangular shape, are put together in different combinations. Five of the modules form a so-called 'pinwheel grid' of larger triangles, the panels. Five of these panels in turn form the constructional unit of the 'mega-panels'.
The architects met the challenge to give buildings that differed so much in terms of their individual briefs unity and yet individuality, too, with these variations. "We wanted to articulate the surfaces because otherwise those buildings would have all been just blank", says Davidson.
The sandstone paving of the 'plaza' reminds the visitor of Australia's vast unknown interior. The architects found the stone in an abandoned stone quarry in the Kimberley Mountains in northwestern Australia. The owner of the quarry is supposed to have traveled to Italy with samples of his stone, where he was accused of having painted its surface himself.

National Gallery of Victoria

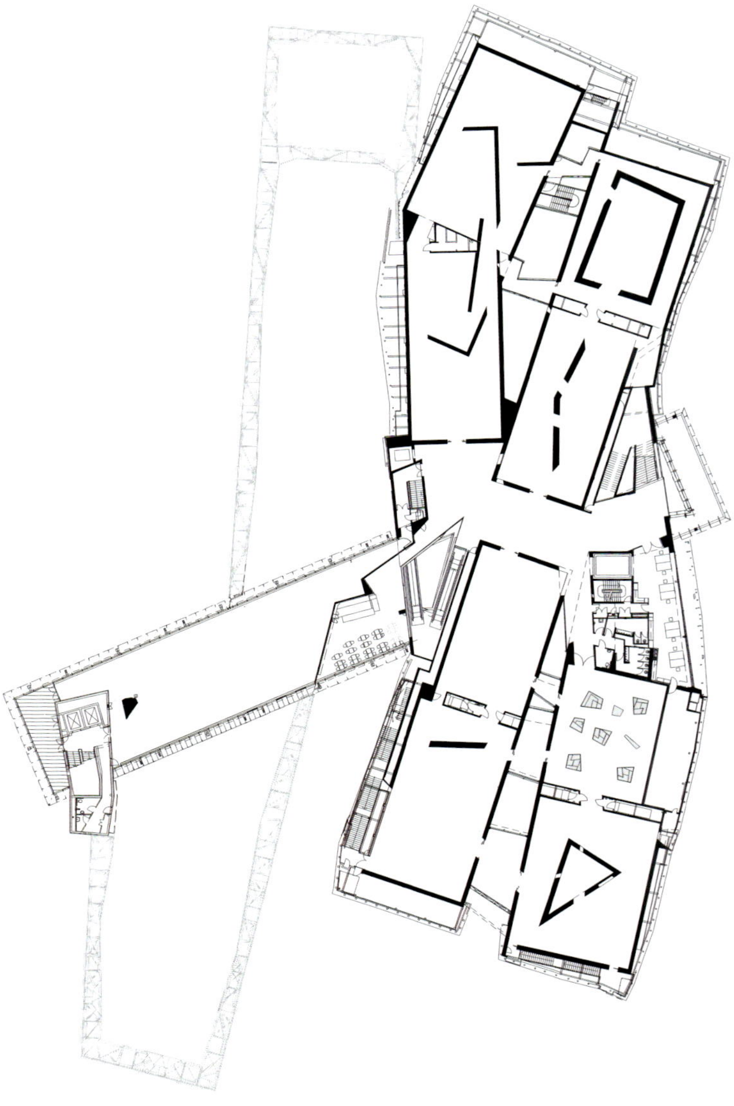

Steine nach Italien gereist sein, wo man ihm vorwarf, diese selbst angemalt zu haben.
In der nordöstlichen Ecke des Square markierten im Entwurf zwei „shards" den Anziehungspunkt und Eingang, das Totem des Federation Square. „Shard" ist das englische Wort für Glassplitter oder Scherbe, in diesem Fall handelt es sich um zwei Türme, die wie Scherben – eine aus Zink, eine aus Glas – in den Himmel ragten.
Aus dem 22 Meter hohen westlichen Glasturm wurde durch eine lang andauernde, kontroversielle Debatte, in deren Verlauf Worte wie Korruption und Arroganz fielen, der wohl politischste acht Meter hohe Glasbaustein der südlichen Hemisphäre. Seitens der Regierung waren Bedenken aufgekommen, dass der *shard* die Sicht auf die dahinter liegende Kathedrale versperren könne. Daniel Libeskind, Jurymitglied des Wettbewerbs, war deshalb eigens nach Melbourne zurückgekehrt, um die Lobby gegen die Kürzung des *shards* zu unterstützen.

Thermo-Labyrinth, urbanes Atrium

Libeskinds Einfluss auf das Projekt gab Kritikern Anlass, die Augenbrauen ein wenig höher zu ziehen. Kritikpunkte betrafen Ähnlichkeiten in der Anordnung der Volumina, die spitzwinklige Geometrie, kluftartigen Zwischenräume und die fraktale Semiologie der Fassade. Hinzu kommt, dass Davidson einer von Libeskinds Mitarbeitern beim Wettbewerb und Design des Jüdischen Museums Berlin war. Wenig verwunderlich also, dass scharfe Zungen Fed Square als seinen „Klon" der südlichen Hemisphäre bezeichneten. Diese Bemerkungen werden von Befürwortern sowie den Architekten selbst mit wenigen Worten zurückgewiesen: „Sie (die Kritiker) könnten einsehen, dass das Gebäude aus einer gedanklichen und praktischen Bewegung mit einem langen Stammbaum hervortritt.(...) Libeskind selbst ist ein Meister dieser Bewegung, aber er ist nicht die Bewegung selbst," schrieb Leon van Schaik, Autor und Professor der RMIT University in Melbourne im Magazin *Monument*.
Als das Budget von AUS$ 128 Mio. ausgeschöpft war, wurden weitere Teile des Entwurfs gestrichen. „Das ursprüngliche Budget war absoluter Unsinn," bemängelt Davidson und erklärt, dass alleine die Kosten für die auf 4.500 Lärm- und Vibrationsschutz-Federn ruhende Plattform beinahe die Hälfte des Budgets verschluckten, welches auch die nachträglich eingeschlossene Nationalgalerie nicht berücksichtigte. Die Kosten betrugen letztendlich AUS$ 450 Mio.
Die Architekten wehrten sich immer wieder gegen Vorwürfe der Extravaganz mit dem Argument, dass Qualität eben nicht ganz billig sei. In vieler Hinsicht zeigten sich Lab als Pioniere in Umwelttechnologie. Das über 140 Meter lange Atrium aus Glas und Stahl, welches sich ungestützt Richtung Norden wölbt, und eine Erweiterung der Gassen Melbournes darstellt, wird durch ein unterirdisches „Thermo-Labyrinth" belüftet, das nachts kühle Luft vom Fluss „einatmet" und damit tagsüber das Atrium bis zu zehn Grad kühler als die Außenluft hält. Photovoltaische Systeme bedienen sich natürlicher Energiequellen.
Eine enge Zusammenarbeit fand zwischen Architekten und den kommissionierten Künstlern statt. Durch einen Wettbewerb waren fünf australische Künstler ausgesucht worden, deren Kunstwerke Teil des Projektes wurden. Die Künstler arbeiteten in den Räumen von Lab und in enger Zusammenarbeit miteinander und den Architekten.
„Viele waren der Ansicht, (das Projekt) benötige keine öffentliche Kunst, denn normalerweise hat diese oft einen kompensatorischen Zweck. Wir dachten, es sei viel interessanter, eine enge Zusammenarbeit herzustellen," erläutert Davidson diese Arbeitsweise.

In the design two 'shards' in the northeastern corner of the Square attract one's attention and signify the entrance; they form the totem of Federation Square. A 'shard' is a splinter of glass, in this case the term is used to describe two towers resembling splinters that project into the sky, one made of zinc, the other of glass.
In the course of a lengthy and highly controversial debate, during which words such as 'corruption' and 'arrogance' were used, out of the 22-metre-high western glass tower there finally emerged what is probably the most political eight-metre-high glass block in the southern hemisphere. Fears had arisen in government circles that the shard could block the view of the cathedral standing behind it. Daniel Libeskind, a member of the competition jury, returned to Melbourne at that time to support the lobby opposed to reducing the height of the shard.

Thermal Labyrinth, Urban Atrium

Libeskind's influence on the project gave critics the opportunity to raise their eyebrows a little higher. The points of criticism included similarities in the layout of the volumes, the acute angled geometry, fissure like intermediate spaces and the fractal semiology of the facades. In addition there was also the fact that Davidson had worked for Libeskind on the competition and design for the Jewish Museum in Berlin. Hardly surprising therefore that sharp tongues described Fed Square as that building's clone in the southern hemisphere. These remarks were refuted by supporters of the project as well as by the architects in just a few words: "They (the critics) could more usefully acknowledge that the building emerges from a school of thought and practice with a long lineage. (...) Libeskind himself is a master of this school, but he is not the school", wrote Leon van Schaik, author and professor at RMIT University in Melbourne, in the magazine *Monument*.
When the original budget of AUS $128 million was exhausted parts of the original design were cancelled. "The original budget was absolute nonsense", criticises Davidson and explains that the cost alone of the platform resting on 4,500 noise and vibration protection springs swallowed almost half the budget, which did not include the National Gallery, which was later added to the brief. The final cost amounted to AUS $450 million.
The architects repeatedly defended themselves against criticisms of extravagance with the argument that quality is simply not cheap. In many respects the architects revealed themself to be pioneers in the area of environmental technology. The atrium of glass and steel that measures over 140 metres in length and curves without supports northwards represents an extension of the lanes of Melbourne. It is ventilated by means of an underground 'thermal labyrinth' that 'inhales' cool air from the river at night and so keeps the atrium up to 10 degrees cooler than the outside air during the day. Photo-voltaic systems make use of natural sources of energy. Close collaboration took place between the architects and the commissioned artists. In a competition five Australian artists were chosen whose art works were to be incorporated in the project. These artists worked in the rooms of Lab, in close collaboration with each other and with the architects.
"Many people said (the project) did not need public art, because maybe conventionally it is used in a compensatory way. We thought it was much more interesting to work in a collaborative way", says Davidson in explaining this method of working.

National Gallery of Victoria

Triumph oder Travestie

Federation Square eröffnete 18 Monate später als vorgesehen. Wie der Zusammenschluss der verschiedenen Staaten Australiens soll dieser neue Teil der Stadt verschiedene Elemente zu einem Ganzen verbinden. Auch in der Gründung der Föderation gab es Verspätungen, Hindernisse und Kritiker der neuen Ordnung. Die Architekten haben durch das Design, die Wahl der Materialien und die Nutzung der natürlichen Energieressourcen eine Architektur geschaffen, die sich ihrer Umgebung anpasst. Dabei soll der Federation Square nicht die Formensprache einer zukünftigen Architektur sprechen, sondern in seine Zukunft hineinwachsen.

Um dies zu realisieren, sind die Architekten Risiken eingegangen, deren Unkalkulierbarkeit Kritik nicht ausschließen konnte. „Hätte ich garantieren können, dass Federation Square funktioniert, dann hätte ich mich nicht genug angestrengt," betont Davidson.

Der Federation Square ist kein Monument wie das Opernhaus Sydneys, aber Davidson betont, dass das auch niemals die Intention war, sondern die Schaffung eines lebenden, atmenden Platzes, der den Bürgern die freie Entscheidung zu seiner Nutzung lassen kann.

Noch immer sind die Meinungen der Fachwelt gespalten und die fertig gestellte Bauten sind nicht ganz der „director's cut". Ob Triumph oder Travestie – die Architekten atmen auf, dass nun das Urteil darüber jenen überlassen bleibt, die davon profitieren sollen: der Öffentlichkeit.

Triumph Or Travesty

Federation Square opened 18 months later than originally planned. Like the union of the different Australian states this new part of the city is intended to combine different elements to form a new whole. In founding the Federation there were also delays, obstacles and criticism of the new order. Through their design, choice of materials and their use of natural sources of energy, the architects have created an architecture that adapts itself to its surroundings. Whereby Federation Square is not intended to speak the formal language of a future architecture but to grow into its future.

To accomplish this the architects took risks that were incalculable and therefore could not avoid attracting criticism.

"If I could guarantee that Federation Square is going to work, then I haven't tried hard enough", emphasises Davidson.

Federation Square is not a monument in the sense of the Sydney Opera House, but Davidson stresses that this was never the intention. Instead the aim was to create a living, breathing open space that leaves it up to the citizens to choose how it should be used.

The opinions of the experts are still divided, and the completed buildings are not entirely the 'director's cut'. Whether triumph or travesty, the architects can breathe a sigh of relief that the final judgment is left to those intended to profit from it: the public.

Raimund Abraham

Raimund Abraham

The Austrian Cultural Forum, New York

Photos Robert Polidori, David Sundberg
Text Matthias Boeckl

Kultur, Magie, Energie

Nach dem 11. September 2001 muss sich New York die Frage stellen, ob die extreme Konzentration wichtiger administrativer und ökonomischer Strukturen die einzige Möglichkeit ist, den Wert dieser Stadt zu erhalten. Im Bezirk Midtown versucht nun – als aktuelle Gegenthese – eine kleine exotische Pflanze im dichten Wald mächtiger Büroriesen die Energien des kulturellen Stadtumbaus zur Wirkung zu bringen. Ihre nahezu ausschließliche Konzentration auf formale Qualitäten ist jedoch kein Luxus: Die Reaktionen zeigen, dass New York den Bau braucht und schätzt. Für diese Zukunft versprechende Wirkung ist fast kein Preis zu hoch.

Culture, Magic, Energy

Since the 11th September 2001 New York has had to confront the question whether the extreme concentration of important administrative and economic structures is the only way to preserve the significance of the city. At present in the Midtown district, as a kind of counter thesis, a small exotic plant is attempting to introduce the energy of cultural urban renovation into the dense forest of mighty giant office blocks. Its almost exclusive concentration on formal qualities does not represent a luxury. The reactions show that New York both needs this building and appreciates it. Almost no price is too high for an effect that promises such a future.

The Berkshire

NYC
fashion
LA GRENOUILLE
FedEx
FedEx

New York war einst der Inbegriff der modernen Stadt – und ist heute fast das Gegenbild einer sinnvoll organisierten und lebenswerten Stadt geworden. Die Stadttransformation hat aber nun auch den Nabel der westlichen Welt „befallen": Zunehmend wandelt sich das Profil New Yorks von dem einer kapitalen Wirtschaftsmetropole zu einer Kultur- und Bildungsstadt. Große Firmen wandern wegen der unzeitgemäßen Bau- und Verkehrssituation schon seit langem mit ihren Logistikstrukturen in den langgezogenen quasi-urbanen Streifen zwischen Boston und Philadelphia aus, nach der Zerstörung des World Trade Centers nun auch vermehrt mit ihren Administrationen. Was übrigbleibt, sind die weniger mobilen Einwohner und die gewaltige ererbte Bau-Hardware. Wegen der riesigen Maßstäbe und der closed-shop-Politik des lokalen Developer-Establishments verläuft der Umbauprozess langsam, wird aber in Manhattan just in jenen Zonen spürbar, wo eine „schwache" Einwohner- und Gebäudestruktur der Expansion einer neuen urbanen Kultur des anspruchsvollen „Lifestyle" weichen muss. Die Stadtverwaltung hat alle Hände voll damit zu tun, die Folgen dieses und vieler anderer in New York wütender Desintegrationsprozesse zu regulieren.

Prestige Kultur

In diesem Ambiente nahezu ungebändigter, in ursprünglicher Reinform auftretender Urbanisierungskräfte war daher die Entscheidung einer der vielen in der Weltmetropole um Aufmerksamkeit ringenden Nationen, für die eigene Profilierung konsequenterweise auch das Hauptmerkmal der eigenen Präsenz am Ort zu radikalisieren, die einzig Richtige: Schon in Zeiten der verklärten k.u.k. Monarchie war die Kunst der wichtigste Exportartikel Österreichs. Joseph Urban etwa wurde nach seiner ökonomisch bedingten Flucht aus Wien 1911 bis zu seinem Tod 1933 zum führenden Theater- und Opernausstatter New Yorks – er hinterließ ihr das 1930 errichtete Gebäude der noch heute florierenden New School for Social Research in der zwölften Straße. Auch der 1926 in die Stadt gekommene Architekturvisionär Friedrich Kiesler baute für die Kultur und entwarf 1928 ein avantgardistisches Multiprojektionskino an der achten Straße. Und die aus Österreich vertriebenen Architekten der Jahre 1938 bis 1945 erinnern baulich durch die Buttinger Library, die Felix Augenfeld 1956 für die sozial engagierte Millionärin Muriel Gardiner-Buttinger errichtete, an ihr Schicksal. Alle diese Bauten brachten – wie nun auch das Kulturforum von Raimund Abraham, der nach seiner Ankunft in New York 1964 im Atelier von Friedrich Kiesler gelebt und so den direkten Kontakt zur Exiltradition hergestellt hatte – einen experimentellen, zumindest unkonventionellen Bauaspekt in ihr jeweiliges Quartier. Mit der fortschreitenden Transformation der Stadt gewinnt nun logischerweise auch die Kompetenz „Kulturbau" an Prestige.

New York was once the definition of a modern city and today has become almost the antithesis of a sensibly organised city and worthwhile place to live. Yet urban transformation has now 'attacked' the navel of the western world. Increasingly New York's profile is changing from that of a capitalist business metropolis to a city based on culture and education. On account of the outdated building and traffic situations large companies have for some time been moving their logistics departments into the long, quasi-urban strip extending between Boston and Philadelphia. Following the destruction of the World Trade Center the administration departments are now following suit. Left behind are the less mobile inhabitants and an immense amount of inherited building hardware. Due to the gigantic scale and the 'closed shop' policies of the local developer establishment the renewal process is taking place sluggishly, in Manhattan it is evident primarily in those zones where a 'weak' resident and building structure must make way for the expansion of the new urban culture based on an ambitious lifestyle. The municipal administration has its hands full trying to regulate the consequences of this and other processes of disintegration currently rampant in New York.

Prestige Culture

In this ambience where almost uncontrolled urbanisation forces appear in their original, unadulterated form the decision of one of the nations that jostle for attention in this world capital to radically treat the main symbol of its local presence in order to establish its profile was the only possible one. Even during the time of the glorified Hapsburg monarchy art was Austria's most important export. After leaving Vienna in 1911, driven by economic need, Joseph Urban was, until his death in 1933, one of New York's leading theatre and opera designers. He left the city the building for the still flourishing New School for Social Research on 12th Street, erected in 1930. The architecture visionary Friedrich Kiesler, who arrived in the city in 1926 built primarily in the area of culture and in 1928 designed an avant-garde multi projection cinema on 8th Street. The fate of the architects forced to leave Austria between 1938 and 1945 is recalled architecturally in the Buttinger Library which Felix Augenfeld built for the socially committed millionaire Muriel Gardiner Buttinger. After his arrival in New York in 1964 Raimund Abraham lived in Friedrich Kiesler's atelier and thus established a direct contact to the exile tradition. His culture forum and the buildings mentioned above brought an experimental, certainly unconventional aspect of building to their respective districts. With the progressive transformation of the city it is only logical that the cultural building sector should become increasingly prestigious.

Zugang von der Lobby zu den Veranstaltungsräumen
Access from the lobby to the multipurpose spaces

Zweierlei Geschichten

Das Spannende an der Geschichte „Fremde bauen in der Stadt" ist, dass sie stets in zwei Teile zerfällt, nämlich in das auswärts geschriebene Kapitel der Projektentwicklung und das vor Ort ausgetragene Drama der Rezeption des Baus. Gemeinsam mit dem Wiener MuseumsQuartier ist das New Yorker Kulturforum Teil einer Kulturbauoffensive der österreichischen Regierung. Ihre Bilanz (15 Jahre vom ersten Wettbewerb bis zur Eröffnung eines der größten Museumskomplexe Europas sowie 10 Jahre vom Wettbewerb bis zur Eröffnung der neuesten New Yorker „Landmark") kann sich im internationalen Kontext trotz der ebenso permanenten wie unangebrachten österreichischen Selbstzweifel und trotz des immer wieder versuchten Boykotts des New Yorker Baus durch die wechselnden Wiener Finanzminister durchaus sehen lassen. Kein anderes europäisches Land mit der Ausnahme Frankreichs investierte in den letzten Jahren staatsoffiziell und derart intensiv in nationale Kulturprojekte internationalen Anspruchs. Und dies gelang trotz dauernder Intrigen diverser populistischer Politiker, die in dieser Sache jedoch wegen des offensichtlich doch vorhandenen Grundkonsenses der „Kulturnation" glücklicherweise nicht nachhaltig punkten konnten.

Two Kinds Of Stories

The interesting thing about the story "foreigners build in the city" is that it always separated into two parts: the chapter on the project development that is written in the native country and the drama of the building's reception which takes place on location. Together with the Vienna Museumsquartier the New York Culture Forum is part of an offensive launched by the Austrian government in the area of cultural buildings. The balance (15 years from the first competition to the opening of one of Europe's largest museum complexes and 10 years from competition to the opening of one of New York's newest landmarks) is acceptable in the international context despite the persistent and equally inappropriate self-doubts of the Austrians and despite repeated attempts by various Austrian Finance Ministers to boycott the New York building. With the exception of France no other European country has in recent years invested officially and so intensively in national cultural projects with international ambitions. These efforts succeeded despite continuous intrigues by various populist politicians who happily, due to the apparent existence of a basic consensus within the "culture nation", failed to score sustainable successes.

Veranstaltungsflächen im Souterrain, geprägt von Treppen und Dachdurchblick
Multipurpose surfaces in the basement, characterised by the stairwell and a view through the roof

1992 lobte das Wiener Außenministerium, das in einem alten New Yorker Townhouse (mit bestmöglicher Midtown-Lage einen Block neben dem Museum of Modern Art) schon seit 1958 ein Kulturinstitut betrieben hatte, einen baukünstlerischen Wettbewerb für einen Neubau am gleichen luxuriösen, aber sehr beengten Grundstück aus. Die Initiatoren waren Peter Marboe und der derzeitige MuseumsQuartier-Direktor Wolfgang Waldner. Zugelassen waren alle in Österreich geborenen Architekten (daher: „baukünstlerisch") ohne Rücksicht auf Status und Ort ihrer Praxis. Aus den 226 von einer Jury unter dem Columbia-Professor Kenneth Frampton beurteilten Einreichungen wurde mit dem Projekt des 1933 in Lienz in Osttirol geborenen und seit 1971 an der New Yorker Cooper Union School lehrenden Raimund Abraham just jene Variante als Siegerprojekt gekürt, die in ihrer Verbindung einer bestimmten architektonischen Haltung (nämlich kompromisslos Kunst zu machen) mit einer bestimmten Biographie (Karriere als bekannter Avantgardist am Ort) ideale Voraussetzungen bot – ein Beweis mehr für die Sinnhaftigkeit derartiger Verfahren. Auf den Plätzen zwei und drei landeten Hans Hollein und Georg Pendl/Elisabeth Senn. Wie ernst man dieses Verfahren zurecht nahm, beweist eine 223 Seiten starke

In 1992 the Viennese Foreign Ministry, which since 1958 had run a cultural institute in an old New York townhouse in the best imaginable midtown location, a block away from the Museum of Modern Art set up an artist-architect competition for a new building on the same luxurious but extremely tight site. The initiators were Peter Marboe and the present director of the Museumsquartier, Wolfgang Waldner. All architects born in Austria were eligible (hence the use of the term artist-architect) without taking into account the status or location of their practice. From the 226 entries examined by a jury under the chairmanship of Columbia professor Kenneth Frampton the project by Raimund Abraham, born in 1933 in Lienz in East Tyrol and a teacher at the New York Cooper Union School since 1971, was selected as winner. This project that connected a specific architectural approach (an uncompromising intention to make art) with a specific biography (a career as a local member of the avant-garde) offered ideal preconditions – a further proof of the sense of such competition processes. Hans Hollein and Georg Pendl/Elisabeth Senn were awarded second and third places respectively. How seriously this process was taken is proven by a 223 page detailed

detaillierte Dokumentation in Buchform, die das Außenministerium danach als Katalog einer Wanderausstellung mit Wortspenden der gesamten österreichischen Architekturschreiberzunft herausgab.
Abraham war vorher vor allem als nicht bauender Architekt bekannt, seine poetischen Architekturzeichnungen sind legendär und in seinem Buch „[un]built" (Springer Verlag, 1996) publiziert.

Avantgarde und Stadt

In der Person Abrahams vermischen sich die österreichischen und die New Yorker Blicke auf das Projekt. Denn in Amerika repräsentiert er seit seiner gemeinsam mit Walter Pichler und Hans Hollein bestrittenen, von Arthur Drexler kuratierten Ausstellung visionärer Entwürfe im New Yorker Museum of Modern Art 1967 jenen Bereich akademischer Kreativität, der in den USA kaum jemals in reale Baugeschäfte kommt. Anders als in Mitteleuropa, wo in der Regel die führenden bauenden Architekten verantwortungsbewusst (oder aus Gewohnheit) auch an Architekturschulen lehren, sind Architekturlehre und Baurealität in den USA strikt geschiedene Bereiche, ja die Architektur selbst nicht einmal unbedingt als kulturelle Disziplin anerkannt. Die fast durchgehende kommerzielle Orientierung wird nur in wenigen prestigeträchtigen Privatbauten und Museumsstiftungen durchbrochen. Aus der Sicht der amerikanischen Publizistik machte es daher viel Sinn, die europäische Auffassung von Baukunst durch einen in den USA akademisch etablierten Europäer zu demonstrieren. Da spielte es auch keine Rolle, dass das künstlerisch-akademische Image des nicht-bauenden Architekten Abraham seinen Bau-Ambitionen durch realisierte Wohn- und Geschäftshäuser in Berlin, Wien und Lienz nicht ganz standhalten kann. Wie in den gezeichneten Idealentwürfen dominieren in diesen Bauten die Wirkungen einer rigiden Geometrie und die archaisierende Auffassung vom Bauen als einem magischen Akt. Schon 1963 hatte Abraham diesen Empfindungen in einem eigenen Buch über „Elementare Architektur" (traditionelle landwirtschaftliche Bauten in den Alpen) nachgespürt. Es erschien übrigens noch vor dem ähnlich orientierten, Buch „Architecture without Architects" (1965) des bereits vor 1938 aus Österreich über Umwege in die USA emigrierten Bernard Rudofsky.

documentation in book form which the Foreign Ministry later published as a catalogue to a travelling exhibition with contributions by the entire corps of Austrian architecture journalists.
Abraham was previously known primarily as a non-building architect, his poetic architectural drawings, which are legendary, are published in his book *[un]built* (Springer Verlag, 1996)

The Avant-Garde And The City

Abraham's person combines both the Austrian and New York view of the project. In America, ever since his joint exhibition with Walter Pichler and Hans Hollein, curated by Arthur Drexler and held in the New York Museum of Modern Art in 1967, Abraham has represented an area of academic creativity that, in the USA, rarely extends as far as the real business of building. In contrast to Central Europe where, as a rule, the leading (building) architects also teach in the schools of architecture (either from a sense of responsibility or tradition), in the USA the teaching of architecture and the reality of building are strictly divorced from each other. Indeed architecture itself is not always recognised as a cultural discipline. This almost completely commercial orientation is abandoned only in the case of a few prestigious private buildings and museum foundations. From the viewpoint of American publications it therefore makes sense to demonstrate the European understanding of the art of building using a European architect academically established in the USA. It is irrelevant that, in fact, the artistic-academic image of Abraham, the architect who does not build, does not quite fit with the reality of his ambition to build as evidenced by his completed housing and commercial projects in Berlin, Vienna and Lienz. As in the idealised design drawings these buildings are dominated by the influence of a rigid geometry and an archaic understanding of building as a magical act. In 1963 Abraham had traced such sensations in his own book on *Elementare Architektur: Architectonics* on traditional farm buildings in the Alps, which, incidentally, appeared before *Architecture Without Architects* (1965) which pursues a similar theme and was written by Bernard Rudofsky who emigrated (indirectly) from Austria to the USA in 1938.

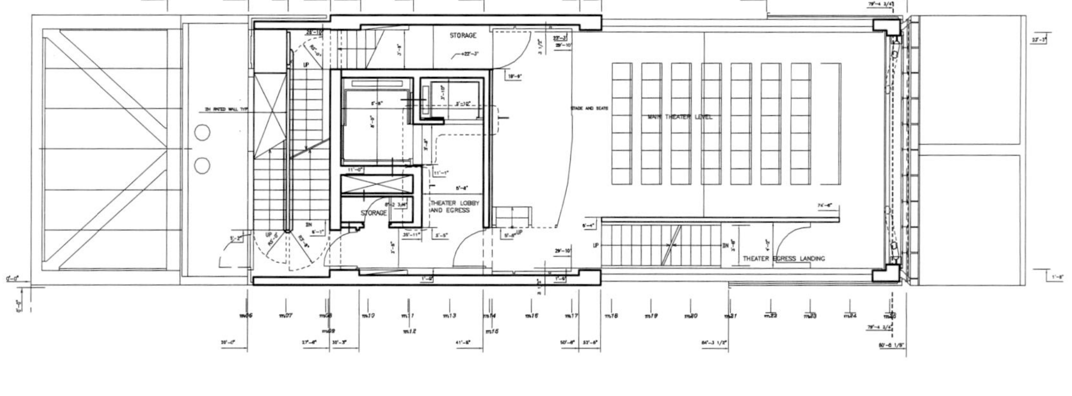

2. Obergeschoß: Auditorium
Level 2: Auditorium

Auditorium mit Empore
Auditorium with gallery

Postmoderne oder antikommerzieller Widerstand?

Lässt sich all das am Bau ablesen? Seine Bedingungen sind extrem, kommen aber der fundamentalistischen Entwurfsphilosophie Abrahams durchaus entgegen, indem sie ebenso extreme Proportionen und Empfindungen provozieren. Der Bauplatz ist ganze sieben Meter breit und 27 Meter tief, 24 dementsprechend schmale und durch die vorgeschriebene Zonierung nach oben immer schmäler werdende Ebenen sind darüber aufgetürmt. Im Wald der riesigen Bürokolosse von Midtown ist das Türmchen nur mit Mühe überhaupt wahrnehmbar, schon seine unmittelbaren, um ein Vielfaches voluminöseren Nachbarn erdrücken das filigrane Gerät visuell. Genau diese Situation ist es aber, die mit einer archaisierenden Strategie gewinnbringend interpretiert werden kann. Enge und Bedrängtheit, Ringen um Autonomie, elementare Empfindungen wie Macht und Eros – das sind exakt jene Themen, die der frühe, ethnologisch gefärbte Fundamentalismus der späteren Postmodernen noch in den 1960er Jahren bearbeitet hatte. Abraham hat sich von dieser Position nie distanziert und konnte daher in der Gunst der historischen Stunde mit einer späten radikalen Geste auf die beschriebene urbanistische Situation reagieren.

Eine durchgehende vertikale Teilung des Gebäudevolumens in einen Treppenschacht, einen Kern und einen vom Architekten treffend „Maske" genannten Fassadenteil erledigt die funktionalen Forderungen gleichermaßen lapidar und demonstrativ. Durch das „Aufsplitten" des Turms gewann Abraham eine allseitig sichtbare plastische Gliederung. Die schräge Fassade ist ein Kunstwerk, das (unter dem hier erstmals im

Postmodern Or Resistance To Commercialism?

Can all this really be read in his building? Its given constraints are extreme but well-suited to Abraham's fundamentalist design philosophy as they provoke in turn extreme proportions and sensations. The site is seven metres wide and 27 metres deep. 24 narrow storeys (which, due to the zoning regulations, grow narrower as the building rises) are stacked on it. In the forest of mighty office colossi in midtown Manhattan one discovers the little tower only with difficulty. Its immediate neighbours, several times as voluminous, threaten to visually squash this filigree appliance. Yet it is precisely this situation that Abraham profitably interprets using a strategy with archaic tendencies. Cramped conditions, constriction, the struggle for autonomy, elemental sensations such as power and eros are exactly the themes which the early fundamentalist (still ethnologically coloured) approach of what later became post modernism dealt with in the 1960s. Abraham never distanced himself from this position and was therefore able to utilise a favourable historic moment and to react with a late, radical gesture to this urban situation.

A continuous vertical separation of the volume into a staircase well, a core and a facade appropriately dubbed "the mask" by the architect fulfil the functional requirements in a way that is equally lapidary and demonstrative. By splitting the tower Abraham achieved a sculptural articulation visible from all sides. This sloping facade is an art work made by the Austrian company GIG, under the supervision of the Viennese client, the "Bundesimmobiliengesellschaft" (Federal Real Estate Company) here for the first time operating in a

Bibliothek: Arbeitsplätze
Library workspaces

Büros
Offices

Ausland agierenden Wiener Bauherrn „Bundesimmobiliengesellschaft") von der österreichischen Firma GIG realisiert wurde und die gesamte Ikonographie der Architektursprache Abrahams vermittelt. Sie ist axialsymmetrisch, zeigt mittig herausstoßende Schwertklingen, vorspringende, scharf umrissene Kuben und metallene Blenden, in die aggressive Keilformen gestanzt sind.

Unscheinbar ist dagegen die Eingangssituation unter der überhängenden Fassaden-„Guillotine", wo der Raum erst im Übergang in die komplexen Innenräume an Dramatik gewinnt. Auf der Eingangsebene sowie auf den Geschoßen darunter und darüber sind in kontinuierlicher Verschränkung die Ausstellungs- und Veranstaltungsräume untergebracht. Darüber türmen sich dann das Auditorium, die Bibliothek, Bürogeschoße und ganz oben auf drei Geschoßen die aus Gründen der Repräsentation hier untergebrachte Direktorswohnung. Das Opfer ist ein doppeltes: Denn erstens könnten die Flächen sinnvoller genutzt werden und zweitens ist eine Wohnung dieser Enge auf mehreren Ebenen für eine Familie mit Kindern eher eine Bürde als eine Würde. In den öffentlichen Zonen der Ausstellungsräume, des intimen, aber konzertfähigen Saales und der Bibliothek ist jeder Raum gleichzeitig auch Erschließungsraum und daher Verkehrszone – ein unvermeidliches Resultat der minimalen Grundstücksfläche, das aber mitunter in ein sinnlich erlebbares Raumkunstwerk umschlägt. Das passiert etwa, wenn man im Untergeschoß durch das Glasdach des rückspringenden Bauteils hinauf in den Wolkenkratzerwald oder ganz oben in der Dachloggia durch eine der dämonischen Öffnungen der „Maske" über ganz Midtown blickt.

foreign country, that communicates the entire iconography of Abraham's architectural language. It is axially symmetrical, along the centre axis sword blades thrust outwards, sharply outlined cubes and metal screens, out of which aggressive wedge shapes have been stamped, project.

The entrance below the projecting facade "guillotine" is unspectacular. It is only in the transition to the complex interior spaces that the space acquires a dramatic quality. On entrance level and the floors directly above and below it the exhibition and public function rooms interlock continuously. Above are the auditorium, library, office levels and, on the top three floors the director's apartment located here for reasons of status and display. The loss is a double one: firstly a more sensible use could have been found for this space and secondly an apartment this narrow on three levels is more of a punishment than a pleasure for a family with children. In the public zones of the exhibition spaces, the salon that is intimate but suitable for concerts and the library each room also serves as its own access and circulation space – an unavoidable result of the minimal site area which, however, is handled to provide a sensually experienced work of spatial art. This is felt for instance when, on the lower floor, one looks upwards through the glass roof of the receding building element into the forest of skyscrapers or, at the very top in the roof loggia, peers through one of the demonic openings of the "mask" across the whole of midtown Manhattan.

Wohnraum
Sitting room

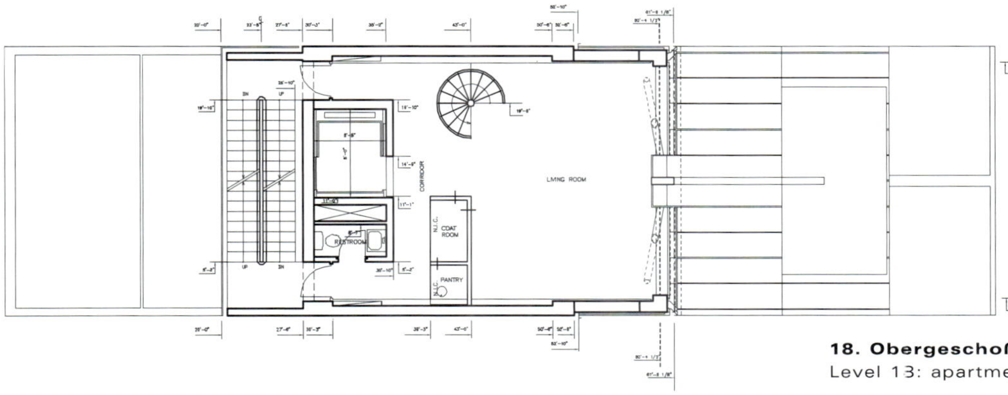

18. Obergeschoß: Wohnung
Level 13: apartment

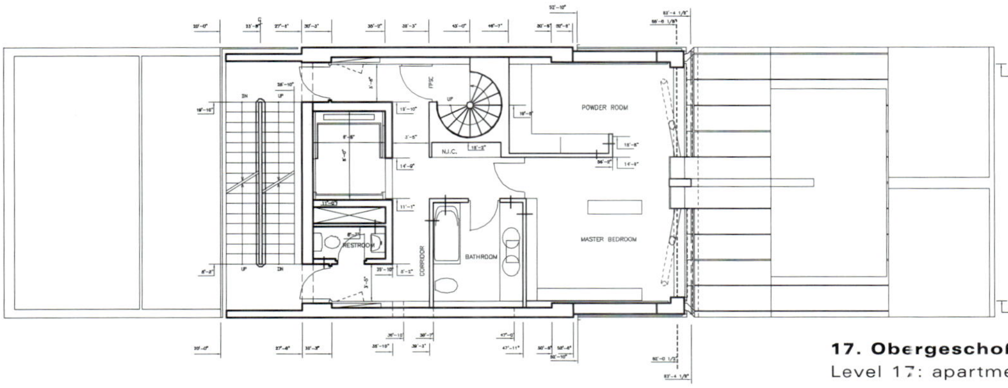

17. Obergeschoß: Wohnung
Level 17: apartment

Fassadenkonstruktion
Construction of the facade

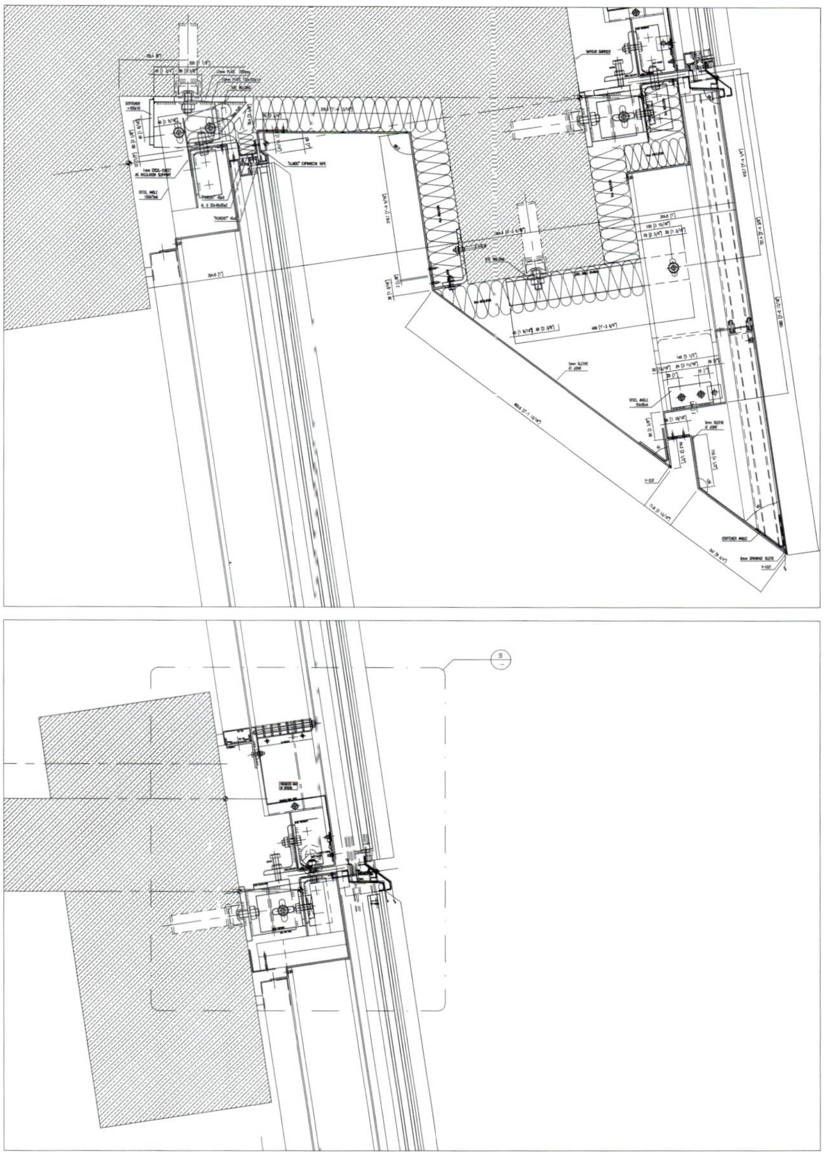

Geschlossene Dachterrasse
Covered roof terrace

Es hat gelohnt

Die Positionen sind eindeutig. Österreich hat sich einen Bau geleistet (den Spott über die überhöhten Kosten sollte man mit Fassung tragen), der in New York als Architektursensation gewertet wird und nun auch auf diesem Niveau – nämlich mit Übergewicht der zeitgenössischen über die historischen Kunstpräsentationen – genutzt werden muss. Die Stadt ist aufnahmebereit und wartet gespannt auf die Energien, die der Bau nun entfalten wird. Die New York Times konstatierte: „Die Lehre des Gebäudes ist eher philosophischer als formaler Art. Es lädt New York dazu ein, sich wieder mit der Unterscheidung zwischen gefälligem Schnickschnack und jener Art von Architektur vertraut zu machen, die etwas in unserem Kopf bewegt." Und das TIME-Magazine sekundierte: „Abrahams kleine Rakete eines Gebäudes zeigt nicht nur, dass die österreichische Kultur lebt, sondern auch, dass die von New York City wiedergeboren werden kann."

Was die Presse bei der Eröffnung aber nicht ansprach: Das Haus bedarf in der Tat einer präzisen Programmierung, um sein Potential ausspielen zu können. Erste Erfahrungen mit der Präsentation „TransModernity" (Architektur von Henke/Schreieck, Jabornegg/Pálffy und Riegler/Riewe) zeigten die erwarteten Defizite des Fassaden-Hauses bei der Ausstellungsnutzung. Und die Veranstaltungen müssen noch viel präziser auf den New Yorker Arbeitsalltag zugeschnitten werden, um nicht, wie bei der Eröffnung und beim Symposion „TransModernity", nur für Österreicher mit viel Zeit zu produzieren, sondern auch Wirkungen über die eigenen Nabelschau-Kreise hinaus zu erzielen.

It Was Worthwhile

The positions are clear. Austria has afforded itself a building (the ridicule over the escalating costs must be borne with countenance) which in New York is regarded as a sensation and which now must be utilised on this level, that is with an emphasis on the presentation of contemporary, not historical art. The city is ready to receive and awaits eagerly the energy which this building will now develop. The New York Times confirmed: "The building's lesson are philosophical rather than formal. It invites New York to reacquaint itself with the distinction between people-pleasing fluff and the kind of architecture that gets inside your head."

And TIME Magazine seconded: "Abraham's little rocket of a building suggests that not only is Austrian culture alive but that maybe New York City's can be born again too."

At the opening there was one thing the press did not refer to: The building in fact needs precise programming in order to exploit its potential. The initial experiences with the presentation TransModernity (architecture by Henke/Schreieck, Jabornegg/Pálffy und Riegler/Riewe) revealed the expected deficiencies of the building and showed that events must be far more closely tailored to the normal New York working day in order to avoid producing only for Austrians with a great deal of spare time as was the case at the opening and at the Transmodernity symposium and in order also to achieve an influence that aims beyond the familiar, self-absorbed circles.

Technische Daten | technical data

Austrian Cultural Forum
New York/U.S.A., 11 East 52nd Street

Bauherr | client **Republik Österreich, Bundesministerium für Auswärtige Angelegenheiten**
Bauträger | developer **BundesImmobilienGesellschaft (BIG), Wien; Hanscomb Inc., New York**
Planung | planning **Raimund Abraham**
Projektkoordination | project coordination **Simone Giostra, Peter Leeb, John Veikos, Jeff Brown**
Mitarbeiter | collaboration **Anders Abraham, Chris Barlieb, Jennifer Whitburn, Catherine Seavitt, Frank Thinius, Mike Derker, Maya Ballen**
Engineering/Consulting **Ove Arup & Partners, New York**
Generalunternehmer | building contractor **Barney/Skanska, New York**
Fassaden | curtain wall **GIG Fassadenbau GmbH, Attnang-Puchheim; Rheinzink, Herzogenburg**
Möblierung/Entwurf | furnishings/design **Atelier Raimund Abraham; Roland Eckl**
Möblierung | furnishings **List, Edlitz**
Lichtplanung/-ausstattung | lighting concept/fittings **Zumtobel Staff**

Grundstücksfläche | site area	196 m²
Nutzfläche netto | net floor area	2.308 m²
Höhe | height	85 m
Geschoße | storeys	24
Wettbewerb | competition	1992
Baubeginn | start of construction	9/1998
Fertigstellung | completion	1/2002

Williams & Tsien

Tod Williams+Billie Tsien

The American Folk Art Museum, New York

Photos Michael Moran
Text Matthias Boeckl

Die Kunst des Widerstands

Einige Monate vor dem Austrian Cultural Forum wurde – nur einen Häuserblock entfernt – mit einem kleinen Museum eine weitere konzentrierte Architekturintervention in das dichte Hochhausrevier von Midtown Manhattan gesetzt. Fast allseitig vom entstehenden Erweiterungsbau des Museum of Modern Art umschlossen, übt das schmale Haus Widerstand sowohl gegen den großen Kulturtanker des MoMA als auch gegen die gebaute Banalität der Bürotürme ringsum.

The Art Of Resistance

Several months before the Austrian Cultural Forum was opened only one block away a small museum made a further concentrated architectural intervention in the dense high-rise quarter of midtown Manhattan. Surrounded on almost all sides by the emerging extension building of the Museum of Modern Art, the narrow building is practising resistance against the great cultural 'tanker' of the MoMA and also against the developmental banality of the office towers around it.

Die Bauaufgabe war jener von Raimund Abrahams Kulturforum durchaus vergleichbar: auf einem extrem engen Grundstück (ca. 13 Meter breit und 30 Meter tief) sollte die Volkskunstkollektion eines ambitionierten Sammlers untergebracht werden. Nicht nur ihr Inhalt – also die bis vor kurzem kaum unter dem in den USA eher traditionalistischen Kunstbegriff gehandelten Alltagsbilder und -objekte von Siedlern aus drei Jahrhunderten amerikanischer Geschichte – sondern auch ihr Anspruch, mitten im Brennpunkt des repräsentativen Kunstgeschehens New Yorks vertreten zu sein, setzte das Vorhaben einem hohen Erwartungsdruck aus. Die Wahl der Architekten Williams und Tsien war für die Ambitionen des Projekts zweifellos die richtige – denn beide repräsentieren führend das „künstlerische" Lager der New Yorker Bauszene. Williams war Mitarbeiter bei Richard Meier und Tsien hatte zuerst als Malerin gearbeitet. Dieses Lager kommt in den USA nur selten zu großen Bauaufträgen und auch Williams/Tsien haben sich ihren Ruf zunächst mit kleinen, exklusiven Bauprojekten der akademischen Welt erworben. Die zahlreichen Campuses privater und öffentlicher Universitäten sind – neben den Einfamilienhäusern – in den USA immer noch eine der wenigen, aber tatsächlich virulenten Experimentierfelder anspruchsvollen Bauens. Wer dort seine innovative Kraft bewiesen hat, der kann sich mit einigem Glück auch im Kulturbau betätigen. Williams und Tsien blicken heute auf eine Handvoll erfolgreicher Arbeiten für Universitäten zurück: Ein Hirnforschungszentrum in La Jolla, ein Kunstzentrum für die Johns Hopkins University in Baltimore und das Schwimmbad der Cranbrook University in Michigan, das in der globalen Architekturpresse bereits gebührend vor allem wegen seiner Materialsensibilität gefeiert wurde.

The assignment may well be compared to that of Raimund Abraham's Cultural Forum: an extremely narrow piece of property (approx. 13 metres wide and 30 metres deep) to be used to accommodate the folk art collection of an ambitious collector. The project was under a great deal of pressure to match expectations: with regard to the content of the collection – namely, the commonplace pictures and objects of settlers covering three centuries of American history, which until recently were hardly regarded as art at all according to the rather traditionalist concept of art prevalent in the US – but also due to the desire to be present at the very hub of New York's representative art business. Considering the project's ambitions there can be no doubt that the choice of the architects Williams and Tsien for the job was the correct one since the two of them are the leading representatives of the 'artistic' camp in New York architecture circles. Williams worked for Richard Meier and Tsien was originally a painter. This camp rarely receives large-scale building contracts in the US and even Williams/Tsien have had to build up their reputation with small and exclusive building projects for the academic world. With the exception of single-family houses, the numerous campuses of private and public universities comprise one of the few (if actually virulent) fields of experiment for more demanding architecture in the USA. Anyone who has shown their innovative power there can, with a bit of luck, also prove themselves in cultural architecture. Williams and Tsien are today able to look back on a handful of successful briefs for universities: a medical research centre in La Jolla, an art centre for the Johns Hopkins University in Baltimore and the swimming pool of Cranbrook University in Michigan, which received very high praise in the global architectural press, above all for its material sensitivity.

Aufwändige Oberflächen (u.); freistehendes Museum vor Realisierung des MOMA-Zubaus
Labour-intensive surfacing (below); freestanding museum prior to the realisation of the MOMA extension

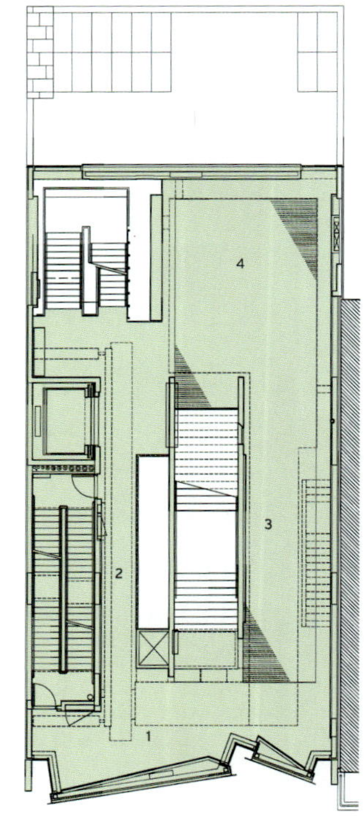

Level 3
1-4 Exhibition

0 10

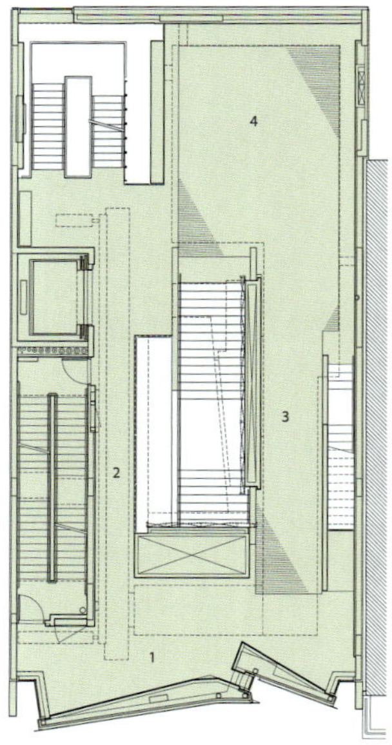

Level 4
1-4 Exhibition

0 10

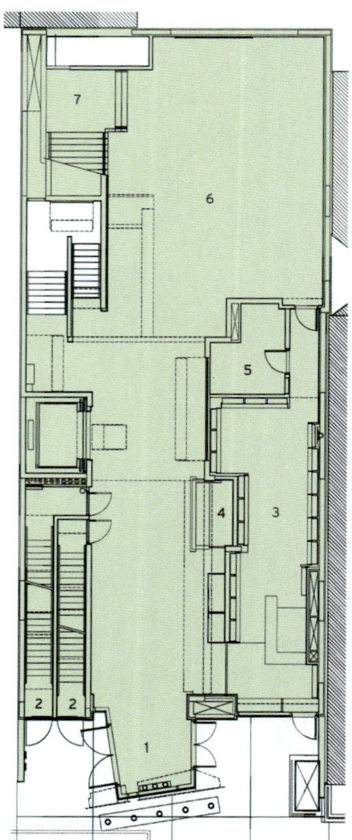

Level 1
1 Lobby
2 Fire Stairs
3 Shop
4 Reception
5 Shop office
6 Hall
7 Stairs

0 10

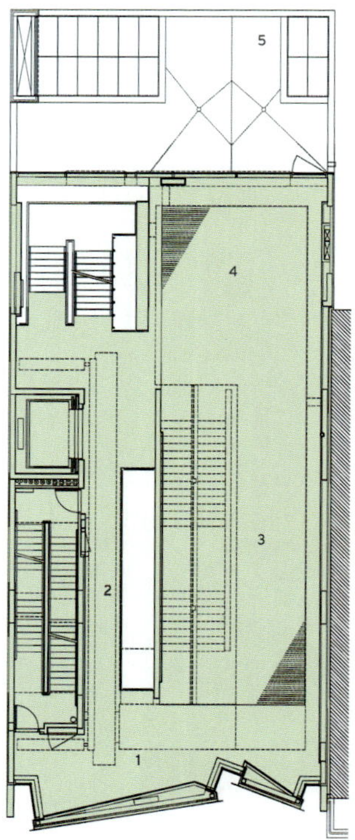

Level 2
1-4 Exhibition
5 Terrace

0 10

Lobby, Reception

Architektur ist schwer

Die Materialdiskussion ist auch der zentrale Aspekt des Folk Art Museums. Die Architekten vertreten in dieser Debatte eine fast fundamentalistische Position. Tod Williams: „Wir glauben, dass Gebäude im allgemeinen mit dem Boden, der Erde verbunden sein sollten. Wir denken Gebäude in den Begriffen von schwer und leicht. Heute gibt es die Vorstellung, dass die Architektur angeblich immer leichter wird. Aber ich glaube das überhaupt nicht. Es ist bloß eine Illusion von Leichtigkeit. Gebäude sind tatsächlich schwer. Ich habe noch nie ein Gebäude gesehen, das ich aufheben könnte.“ Die entwerferische Haltung von Williams und Tsien beschränkt sich aber nicht darauf, die chtonische Gebundenheit des Bauens zu beschwören. Im Gegenteil, denn gerade das Folk Art Museum zeigt, wie raum- und lichthaltig gerade die Mauermasse werden kann, wenn sie gezielt perforiert, verschoben oder gespalten wird. Im Folk Art Museum öffnet sich ein zentraler Canyon, der die Mauermassen von unten nach oben vertikal durchschneidet und der dem Besucher auch eine Art Aufstiegsweg zum Licht suggeriert. Er ist umso wirkungsvoller, als er auf Schritt und Tritt mit subtil bearbeiteten, meist gestockten Sichtbetonflächen, mit authentisch-schwerem Holzboden, eingebauten Holzbänken, Exponatträgern aus Holz und Stahl sowie vielen anderen gewichtigen Materialien konfrontiert wird – all das bei einem geringen Innenraumvolumen, das angesichts der Bauabmessungen überhaupt genutzt werden konnte.

Architecture Is Heavy

The debate about materials also relates to one of the main aspects of the Folk Art Museum. These architects occupy an almost fundamentalist position in this debate. Tod Williams: "We feel that buildings in general should be connected to the ground, the earth. We think of buildings in terms of heavy and light. There is a notion these days that architecture is increasingly becoming lighter. But I don't believe it one bit. It's just an illusion of lightness. Buildings are heavy. I haven't met a building I could lift." However Williams and Tsien's approach as designers is not restricted to invoking the chthonic affinity of architecture. On the contrary, the Folk Art Museum shows how precisely the mass of the walls can contain space and light if they are deliberately perforated, shifted or split. In the Folk Art Museum a central canyon opens up which cuts vertically through the mass of the wall, from below to above, suggesting to the visitor a kind of way up towards the light. This is all the more effective since the visitor is confronted at every step with subtly worked and usually bush-hammered fair-faced concrete areas, authentically heavy wooden flooring, built-in wooden banks and exhibit supports made of wood and steel, as well as many other heavy materials – and all of that within the smallest interior volume that could possibly be used for a building of these dimensions.

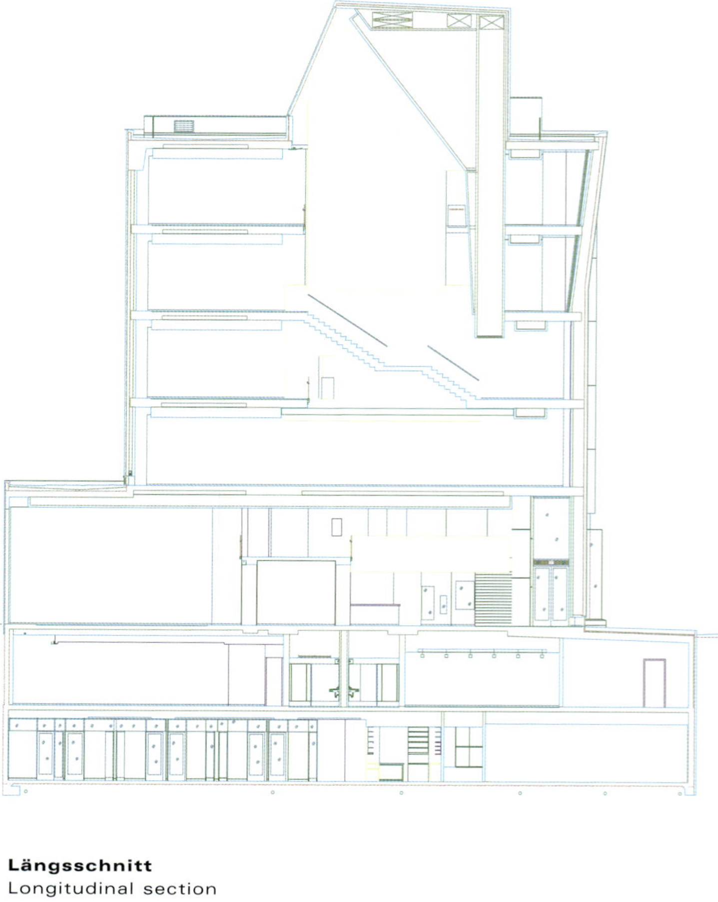

Längsschnitt
Longitudinal section

EXIT

Einbrüche, Durchbrüche

Mit gezielt gesetzten Durchbrüchen wird der Bau jedoch sowohl außen als auch innen der tellurischen Verhaftung immer wieder enthoben. Das beginnt schon an der Fassade. Ursprünglich wollten die Architekten noch „erdhaftere" Materialien als in der letztendlichen Realisierung, wofür etwa Marcel Breuers Betonmassen des in den 1960er Jahren gebauten Whitney Museum of American Art – das von der Zeitschrift „Domus" übrigens als das letzte anspruchsvolle Stück Architektur in Manhattan bezeichnet wurde – als Vorbilder dienten. Williams/Tsien wollten aber nicht nur das Material, sondern auch den Ort selbst zur Sprache kommen lassen: „Unsere erste Idee war es, einen vor Ort gegossenen Betonabdruck der 53. Straße als Fassade zu verwenden. Das wäre ziemlich spannend gewesen, hätte aber den Autoverkehr zu stark behindert. Der Vorstand des Museums hat nicht gerade positiv auf unsere Idee reagiert, die Fassade aus Ortbeton zu machen. Also haben wir uns um andere Materialien umgesehen. Zuerst dachten wir an Aluminium, weil es funkelt und glänzt und billig genug ist, dass man es sich leisten kann. Als wir aber einmal einen Barren Aluminium schmolzen und in eine Form gossen, glänzte es überhaupt nicht mehr. Es wurde richtig stumpf und sah gar nicht mehr wie Aluminium aus. Also sagten wir: ‚Suchen wir doch etwas, das weder Aluminium noch Kupfer ist, sondern vielleicht eine Mischung von beidem.' Nach diversen Recherchen stießen wir schließlich auf Tombasil, das eine Art weiße Bronze ist." Dieses Material ermöglichte es, sowohl die erdhafte Ausstrahlung des Baus an der Fassade zu realisieren, als auch durch metallische Glitzereffekte in den Umraum abzustrahlen – auf jeden Fall eine auffallende Fassade, die im Umfeld der Midtown-Bürohäuser ziemlich eindeutig auf einen Kunstaspekt hinweist.

Break-Ins And Breakthroughs

With deliberately designed breakthroughs the building is time and again raised above its tellurian restrictions both outside and inside, a process that already begins with the facade. Originally the architects wanted materials that were even more 'earthy' than those used in the final implementation, taking as their model Marcel Breuer's use of concrete masses in the Whitney Museum of American Art built in the 1960s – which, incidentally, was described by the journal Domus as the last piece of ambitious architecture in Manhattan. However, Williams/Tsien did not want to let only the material speak for itself, but also the location: "Our first idea was to try to cast a concrete facade directly on the street, to use the imprint of 53rd Street as the facade. That would have been extremely interesting, but rather disruptive to traffic. We didn't get a very positive response from the Folk Art board to the idea of making the facade of cast concrete. So we looked to other materials. We thought of aluminium because aluminium can have some sparkle and shine and it is ordinary enough to be affordable. But once we melted an ingot and then cast it, it was no longer shiny. It became rather dull and didn't look like aluminium. So we decided to look at something that was not aluminium and not copper, but maybe a mix of the two. Through a series of investigations, we learned about Tombasil, which is a form of white bronze."

This material makes it possible to give the building an earthy aura on the facade and to radiate into the environment by means of metallic glittering effects – at any rate, to achieve a conspicuous facade that indicates the artistic aspect of the building rather clearly amidst the surrounding midtown office blocks.

The introverted gesture of the three large facade elements, which are composed of single cast slabs, is ambivalent: on the one hand, such a hermetic approach unmistakably indicates the autonomy and 'otherness' that is being striven for in the building, on the other hand the exposure and lively surface relief that results from the casting promises an interesting interior life. The three wings of the facade, which are folded inwards from the sides like a cardboard box, are blocked to such a degree that they leave only a narrow slit in the middle free, and this serves as a light source for the multi-storey inner hall. Yet in the case of the facade precautions have also been taken should it be totally closed in by the two adjoining neighbouring buildings: for it is not finished flush, but ends so far away from the side edge of the building that the ensuing gap once again gives a view of the interior of the building. This old practice of leaving one's own building free from 'all-too harmonious' immediate neighbours was also used by Otto Wagner in his town houses in Vienna.

Die introvertierte Gestik der drei großen, aus einzelnen gegossenen Platten zusammengesetzten Fassadenteile ist ambivalent: Einmal weist diese Hermetik unmissverständlich auf die angestrebte Autonomie und „Andersartigkeit" des Baus hin, andererseits verspricht sie mit ihrer Öffnung und dem im Guss entstandenen lebhaften Oberflächenrelief ein interessantes Innenleben. Die drei wie bei einer Kartonschachtel von den Seiten nach innen gefalteten Fassadenflügel sind so arretiert, dass sie nur einen schmalen Schlitz in der Mitte freilassen, der als Lichtquelle für die mehrgeschoßige Innenhalle dient. Die Fassade sorgt aber auch schon für den Fall des totalen Einbaus in die beidseits heranstoßenden Nachbarbauten vor: Denn sie ist nicht seitenbündig ausgeführt, sondern endet so weit vor den Seitenkanten des Baus, dass sie im entstehenden Spalt wieder Einblicke ins Gebäudeinnere freigegeben werden. Diesen alten Kunstgriff des „Freistellens" des eigenen Gebäudes bei einer nicht „gleichklingenden" unmittelbaren Nachbarschaft hat schon Otto Wagner in seinen Wiener Stadthäusern angewendet.

Precarious Balances

The interior development is characterised by the spatial restrictions for the course of the museum and its exhibits, as well as the attractive views of the museum environment. Although it remains visible in space, the staircase access has been shifted to the left rear side of the narrow piece of property (in his Cultural Forum, Raimund Abraham uses solely the rear – which in this way becomes 'blinded'). The whole museum is one single spatial continuum arranged according to levels and smaller dividing walls. Yet besides the elegant craftsmanship and noticeable presence of surface materials, it is also the character of the exhibits which determines the spatial experience. Above all it is the painted wooden objects (house signs, toys, tools and household utensils primarily derived from the settlement areas along the East Coast), the textiles and paintings that produce the intimate, moving atmosphere. The architects have attempted to produce a harmony with the settlers' traditions of craftsmanship – and in that they have succeeded.

Prekäre Balancen

Die innenräumliche Entwicklung ist von der Beengtheit des Parcours und den Exponaten bei gleichzeitigen mitreißenden Ausblicken in die Umgebung des Museums geprägt. Die Treppenerschließung ist an die linke hintere Seite der schmalen Grundstücksfläche geschoben (Raimund Abraham setzte sie bei seinem Kulturforum überhaupt an die – damit „erblindete" – Rückseite), steht aber sichtbar im Raum. Das ganze Museum ist ein einziges Raumkontinuum, das durch Ebenen und kleinere Trennwände gegliedert ist. Neben der gediegenen Handwerklichkeit und den sehr präsenten Oberflächenmaterialien ist es aber auch der Charakter der Exponate, der das Raumerlebnis prägt. Vor allem bemalte Holzobjekte (Hausschilder, Spielzeug, Werkzeug und Hausrat vorwiegend aus den Siedlungsgebieten entlang der Ostküste), Textilien und Gemälde sind es, die diese intime, berührende Stimmung erzeugen. Die Architekten haben versucht, einen Gleichklang mit den handwerklichen Traditionen der Siedler herzustellen – und das ist ihnen auch gelungen. Wie in einer Holzschatulle fühlt man sich hier, und der Kontrast zu der am wenigsten handwerklichen, unpersönlichsten und ausgesetztesten Umwelt, den man durch die Blicke aus den schmalen, aber gezielt gesetzten Sehschlitzen auf Midtown Manhattan erfährt, steigert das nur noch. Diese prekäre Balance zwischen der „großen" Welt draußen und der „heilen" ländlichen Welt der ersten Fischer, Händler und Viehzüchter im Land vermittelt sich auch über die letztlich doch keine Gemütlichkeit aufkommen lassenden Wand- und Deckenflächen, die immer wieder das Raumkontinuum durchschneiden. Williams und Tsien lassen kein Missverständnis über die unbedingte Gegenwärtigkeit des Hauses aufkommen.

It feels as if one were in a wooden casket here, a feeling that is only increased by the contrast to the environment outside, which exhibits little craftsmanship, is highly impersonal and extremely exposed, as glimpsed in the views of midtown Manhattan to be obtained through the narrow but deliberately placed slits. This precarious balance between the 'huge' world outside and the 'intact' rural world of the first fishermen, traders and cattle breeders who settled in this land is also conveyed by those areas of wall and ceiling which time and again cut through the spatial continuum and in the final analysis do not allow any sense of cosiness to develop. Williams and Tsien do not allow any misconceptions to arise about the absolutely contemporary nature of this building.

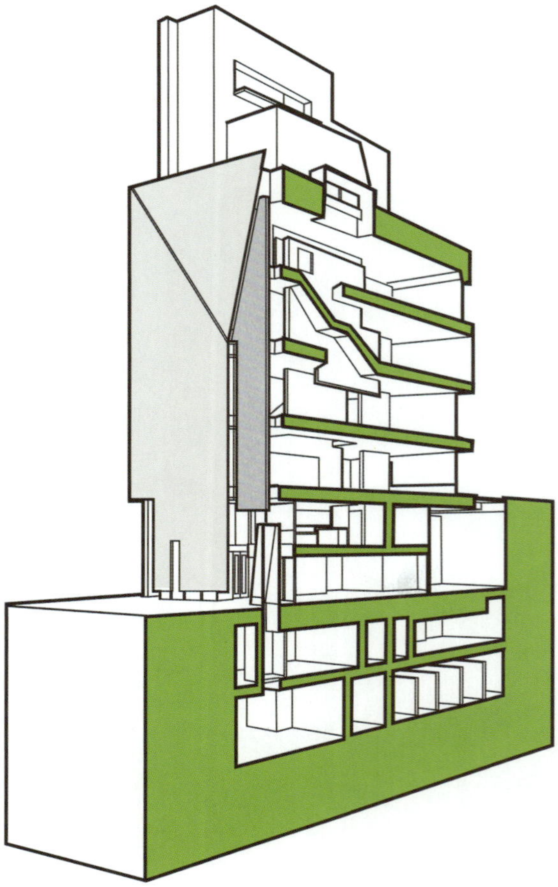

Technische Daten | technical data

American Folk Art Museum
New York/U.S.A., 45 West 53rd Street

Bauherr | client **American Folk Art Museum**
Generalunternehmer | building contractor **Pavarini Construction Company**
Planung | planning **Tod Williams Billie Tsien Architects**
Projektleitung | project manager **Matthew Baird**
Mitarbeiter | assistance **Jennifer Tulley, Philip Ryan, Jonathon Reo, Jennifer Turner, Nina Hollein, Vivian Wang, Hana Kassem, Kyra Clarkson, Andy Kim, William Vincent, Leslie Hanson**
Statik | structural consultant **Severud Associates**
Fassaden | facade **Axis Group Limited**
Elektroinstallationen | electrical services **Commercial Electrical Corp.**
Heizung/Lüftung/Klima | heating/ventilation/air conditioning **React Industries**
Lichtplanung/-ausstattung | lighting concept/fittings **Renfro Design Group**
Aufzug | elevators **Thyssen Dover**

Grundstücksfläche | site area	372 m²
Nutzfläche | floor area	2.694 m²
Planungsbeginn | start of planning	6/1997
Baubeginn | start of construction	10/1999
Fertigstellung | completion	12/2001

Ernst Beneder

Ernst Beneder

Stadtprojekt Waidhofen an der Ybbs, Niederösterreich
Waidhofen An Der Ybbs Urban Project, Lower Austria

Photos Margherita Spiluttini
Text Matthias Boeckl

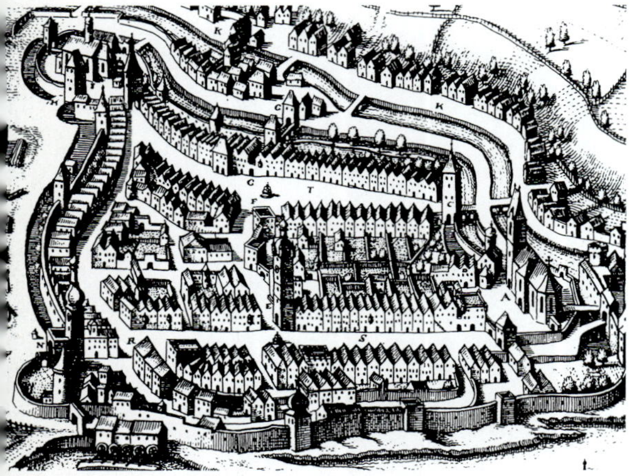

Matthäus Merian: Waidhofen 1649

Sehen – erkennen – bauen

Seit mehr als zehn Jahren arbeitet Ernst Beneder mit und für eine ländliche Kleinstadt an einem exemplarischen Projekt, das die Aufwertung des Lebens in einer alten Kernstadt anstrebt. Wie alle anderen alten Kleinstädte Mitteleuropas ist Waidhofens Innenstadt von Megaattraktoren ringsum bedroht. Ein hoher Identifikationsgrad der Bürger mit ihrer Stadt und die konsequente jahrelange Kommunikationsarbeit des Architekten mit Stadtverwaltung und Einwohnerschaft haben aber ein kleines Wunder vollbracht: Die Kernstadt steigert ihre Lebensqualität und wendet damit die übliche stadtzerstörende Dynamik des Ausrinnens in den suburbanen Ring konsequent ab.

Seeing – Acknowledging – Building

For more than ten years Ernst Beneder has been working for and with a small rural town on an exemplary project that seeks to improve the quality of life in an old town centre. Like all other small old towns in Central Europe the centre of Waidhofen is threatened by the mega-attractors surrounding it. A high level of identification of the citizens with their town and the architect's consistent work over the years with both municipal authorities and residents have led to a small miracle. The quality of life in the town centre has improved, enabling it to consistently deflect the destructive dynamism resulting from Seepage to the suburban ring.

REISEBÜRO

Ernst Beneders Stadtprojekt ist zweifellos exemplarisch – aber auch singulär, denn die Voraussetzungen sind die spezifischen Eigenschaften des Orts: Waidhofen liegt an einem wirtschaftshistorisch bedeutenden Fluss in einem Talkessel am Rande der Alpen. Entlang der Ybbs hatte sich schon seit dem Mittelalter bis in entlegene Winkel hinterer Seitentäler die komplexe Metall- und Holzwirtschaft der sogenannten „Eisenwurzen" entwickelt. So entstanden in Waidhofen als erster größerer Stadt am Wege der Rohstofftransporte (Eisen, Holz) aus den Bergtälern in die großen Verarbeitungszentren des Alpenvorlandes zahlreiche metall- und holzverarbeitende Betriebe. Als inzwischen europaweite „Player" entwickelten sich daraus die Stahlwerke Böhler-Uddeholm und die Büromöbelfabrik Bene. Neben diesen überregional bedeutenden Arbeitgebern und Technologiegeneratoren stehen heute aber auch die mittelständische Wirtschaft der Statutarstadt sowie der Bildungssektor in Hochblüte. Die Integrität war bis in die napoleonische Zeit durch die Besitzer der Stadt, die Bischöfe von Freising in Bayern, garantiert worden. Auf diesen Fundamenten einer kompakten, im Kern mittelalterlichen, in der Bausubstanz meist gotischen bis barocken Stadt ließ sich die Sanierung nach der Transformation des Wirtschaftsgefüges in den vergangenen Jahrzehnten solide gründen. Und der historisch-natürliche Charakter mit spektakulären Blicken über eine Schlucht, einem romantischen Schloss (ehemals im Besitz der Freisinger, dann der Rothschilds, heute der Stadt), den Stadtmauern, Bürgerhäusern, gotischen Kirchen und dem Stadtturm bietet auch (noch nicht ausgeschöpfte) touristische Möglichkeiten.

Ernst Beneder's urban project is without doubt exemplary, but also unique as the preconditions are the specific characteristics of the place. Waidhofen lies on a historically economically significant river in a valley at the edge of the Alps. From the Middle Ages onwards the complex metal and timber economies of the so-called 'Eisenwurzen' developed along the river Ybbs, extending into the far corners of remote side valleys. In Waidhofen, the first larger town along the transport route of the raw materials (iron and timber) from the mountain valleys to the large processing centres of the Alpine foothills, numerous metal and timber processing industries were established. The Böhler Uddeholm steelworks, now a pan-European 'player', and the Bene furniture factory developed out of this situation. Alongside these significant supra-regional employers and technology generators, the medium-size economy of the statutory town and the educational sector are flourishing. Until Napoleonic times the integrity of the place was guaranteed by the owners of the town, the bishops of Freising in Bavaria. The renovation that followed the transformation of the economic infrastructure in recent decades could be soundly based on the foundations of a compact town with a mediaeval core and built substance ranging mostly from the Gothic to Baroque periods. The natural historic character with spectacular views across a ravine, a romantic castle (once owned by the Freising bishops, then by the Rothschilds and today by the town itself) town walls, patricians' houses, Gothic churches and the Stadtturm (town tower) offer tourism possibilities not (as yet) completely exploited.

Infotafeln für die Bürger – realisierte Neugestaltung des Oberen Stadtplatzes
Information boards for town residents – completed redesign of Oberer Stadtplatz

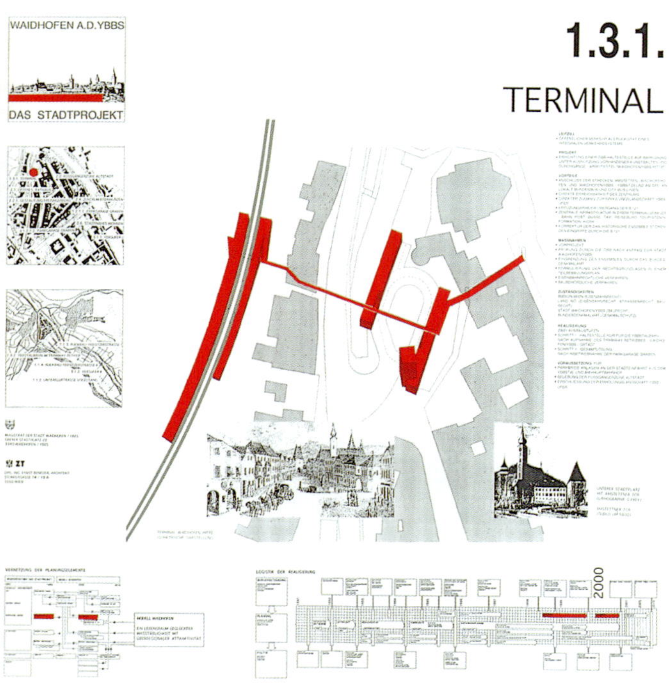

EISEBURO

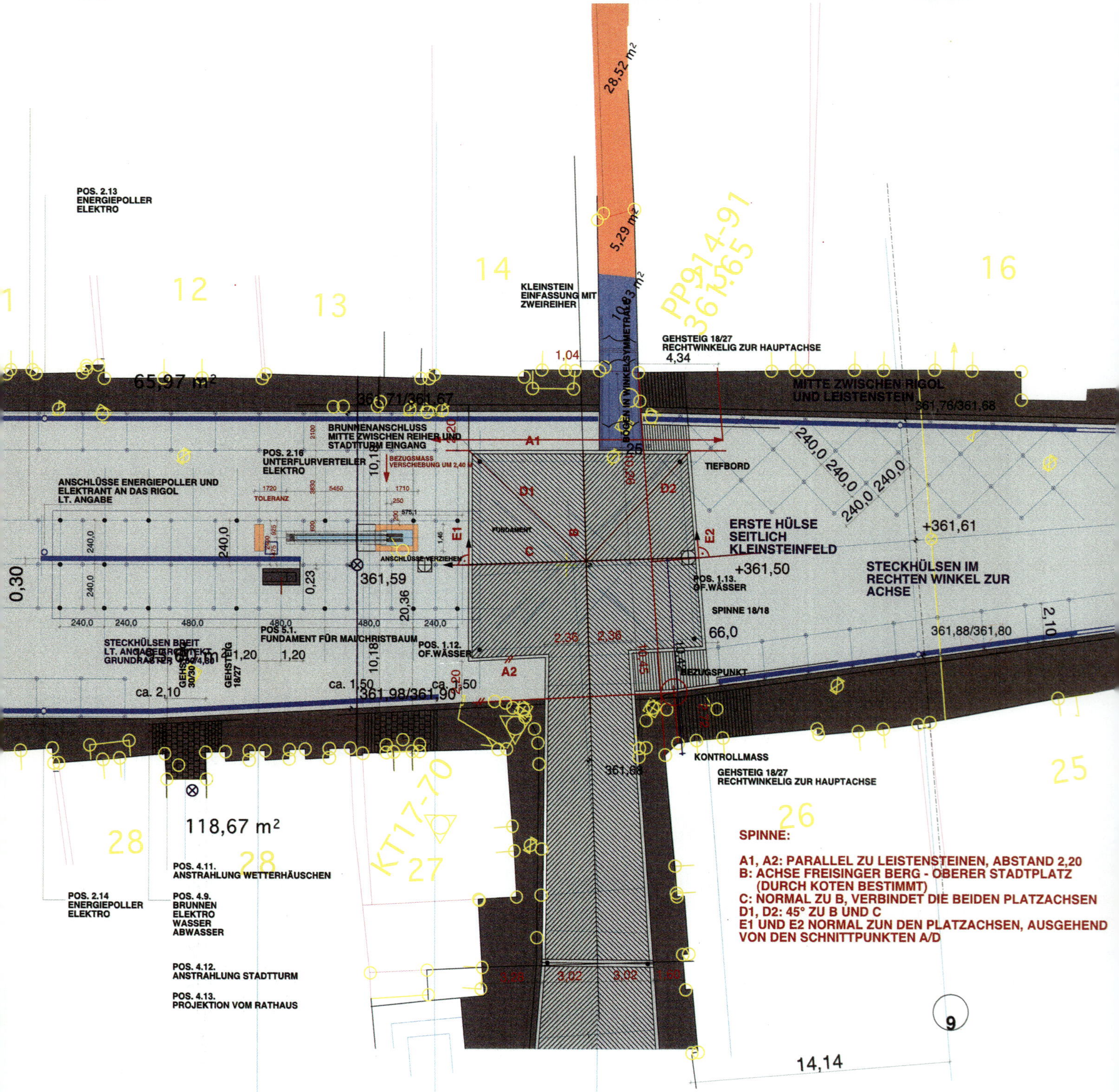

POS. 2.13
ENERGIEPOLLER
ELEKTRO
KLEINSTEIN
EINFASSUNG MIT
ZWEIREIHER
28,52 m²
5,29 m²
GEHSTEIG 18/27
RECHTWINKELIG ZUR HAUPTACHSE
4,34
1,04
65,97 m²
MITTE ZWISCHEN RIGOL
UND LEISTENSTEIN
361,76/361,68
BRUNNENANSCHLUSS
MITTE ZWISCHEN REIHER UND
STADTTURM EINGANG
BÖGEN IN WINKELSYMMETRALE
POS. 2.16
UNTERFLURVERTEILER
ELEKTRO
BEZUGSMASS
VERSCHIEBUNG UM 2,40 M
TOLERANZ
ANSCHLÜSSE ENERGIEPOLLER UND
ELEKTRANT AN DAS RIGOL
LT. ANGABE
TIEFBORD
240,0
ERSTE HÜLSE
SEITLICH
KLEINSTEINFELD
+361,61
+361,50
STECKHÜLSEN IM
RECHTEN WINKEL ZUR
ACHSE
POS. 1.13.
OF.WASSER
SPINNE 18/18
FUNDAMENT
ANSCHLÜSSE VERZIEHEN
361,59
0,30
0,23
20,36
10,18
66,0
361,88/361,80
2,10
POS 5.1.
FUNDAMENT FÜR MAICHRISTBAUM
POS. 1.12.
OF.WASSER
BEZUGSPUNKT
STECKHÜLSEN BREIT
LT. ANGABE
GRUNDRASTER
GEHSTEIG
18/27
1,20
ca. 2,10
ca. 1,50
361,98/361,90
361,68
KONTROLLMASS
GEHSTEIG 18/27
RECHTWINKELIG ZUR HAUPTACHSE
118,67 m²
POS. 4.11.
ANSTRAHLUNG WETTERHÄUSCHEN
POS. 2.14
ENERGIEPOLLER
ELEKTRO
POS. 4.9.
BRUNNEN
ELEKTRO
WASSER
ABWASSER
POS. 4.12.
ANSTRAHLUNG STADTTURM
POS. 4.13.
PROJEKTION VOM RATHAUS
SPINNE:
A1, A2: PARALLEL ZU LEISTENSTEINEN, ABSTAND 2,20
B: ACHSE FREISINGER BERG - OBERER STADTPLATZ
(DURCH KOTEN BESTIMMT)
C: NORMAL ZU B, VERBINDET DIE BEIDEN PLATZACHSEN
D1, D2: 45° ZU B UND C
E1 UND E2 NORMAL ZUN DEN PLATZACHSEN, AUSGEHEND
VON DEN SCHNITTPUNKTEN A/D
3,02
9
14,14

Stadt-Transformation

Die Geschichte begann mit einem Wettbewerb zu einem Gestaltungs- und Verkehrskonzept (beide Bereiche hatten sich vorher unbefriedigend entwickelt), den die Stadtverwaltung 1991 ausschrieb. Ernst Beneder, ein Sohn der Stadt, der schon als Jugendlicher durch Anfertigung einer flächendeckenden Fotodokumentation der Bauten eine intensive Einfühlung in Formen und Funktionsweisen der kleinstädtischen Strukturen erworben und in einem Japanaufenthalt auch andere Stadtkulturen kennengelernt hatte, gewann das Verfahren – und damit die Möglichkeit, in enger Interaktion mit Verwaltung und Bürgern der 10.000-Einwohner-Gemeinde die einzelnen Sanierungsaufgaben und Aufwertungsideen in einem Leitprojekt zu definieren. Und sie so zu streuen und kommunizieren (eine notwendige, aber zeitaufwendige und vor allem unbezahlte Tätigkeit!), dass eine gemeinschaftliche Dynamik und ein konstruktives allgemeines Interesse an architektonischen und urbanistischen Fragen entstand. Beneders Strategie war es, die einzelnen Interventionen nacheinander anzugehen und damit auch als offenen Prozess statt als hermetisches Projekt mit von vornherein feststehendem Fertigstellungstermin zu definieren.

Urban Transformation

The story starts with a competition for a design and traffic concept (both of which areas had developed in an unsatisfactory manner) that the municipal authorities set up in 1991. Ernst Beneder is a native of the town. When still young he acquired an extensive introduction to the forms and methods of functioning of small town structures by making an extensive photographic documentation of the place, during a stay in Japan he become familiar with very different urban cultures. Beneder won this competition and so was given the possibility of defining the individual renovation projects and ideas for improvement in the framework of a master project while working in close collaboration with the residents and the town administration. He was able to spread and communicate these ideas (a necessary but time-consuming and unpaid activity), so establishing a social dynamic and a constructive general interest in issues of architecture and town planning. A specific strategy was to carry out the individual interventions one after the other, defining them as an open process instead of as hermetic projects with predetermined completion deadlines.

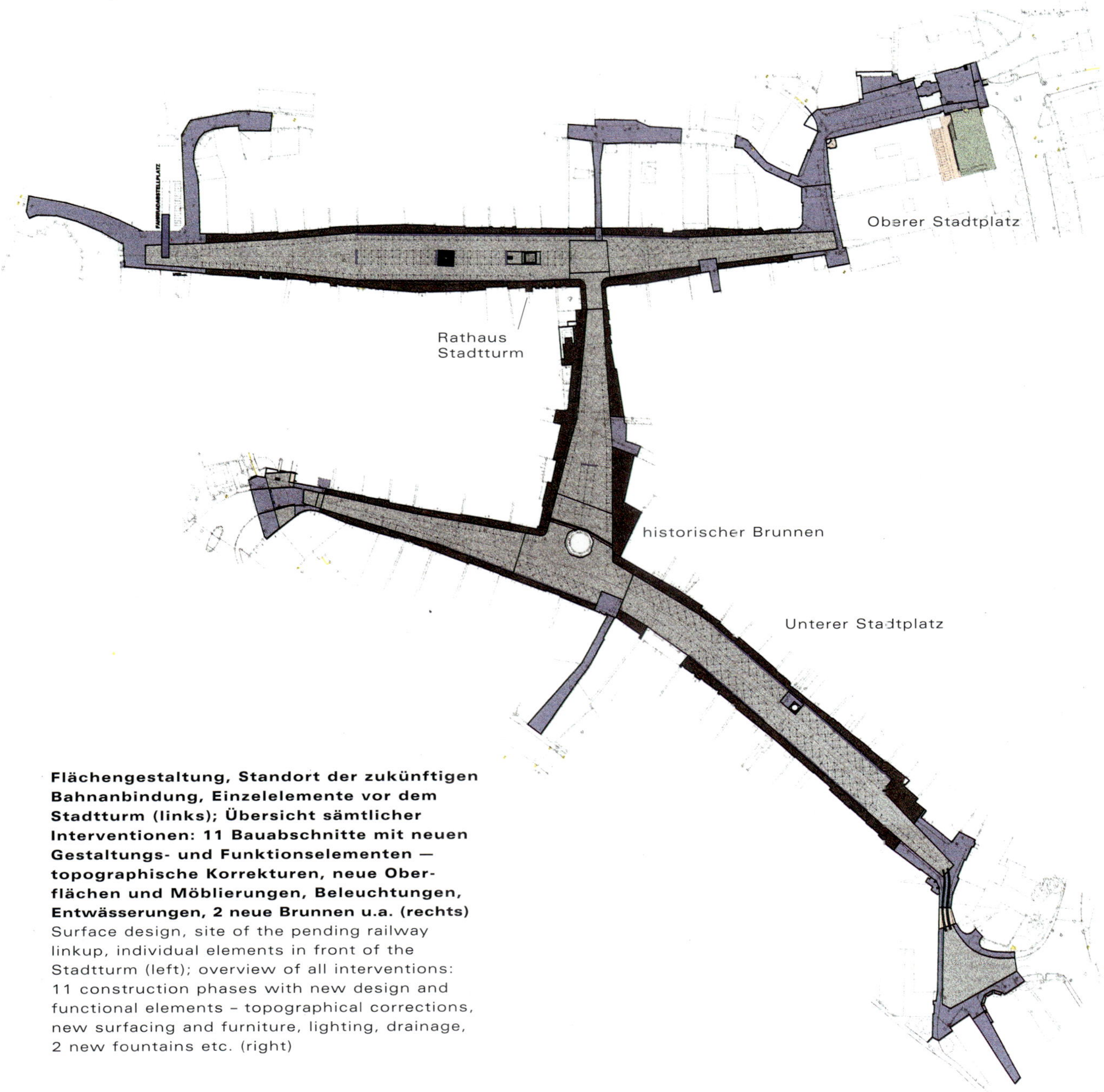

Flächengestaltung, Standort der zukünftigen Bahnanbindung, Einzelelemente vor dem Stadtturm (links); Übersicht sämtlicher Interventionen: 11 Bauabschnitte mit neuen Gestaltungs- und Funktionselementen — topographische Korrekturen, neue Oberflächen und Möblierungen, Beleuchtungen, Entwässerungen, 2 neue Brunnen u.a. (rechts)
Surface design, site of the pending railway linkup, individual elements in front of the Stadtturm (left); overview of all interventions: 11 construction phases with new design and functional elements - topographical corrections, new surfacing and furniture, lighting, drainage, 2 new fountains etc. (right)

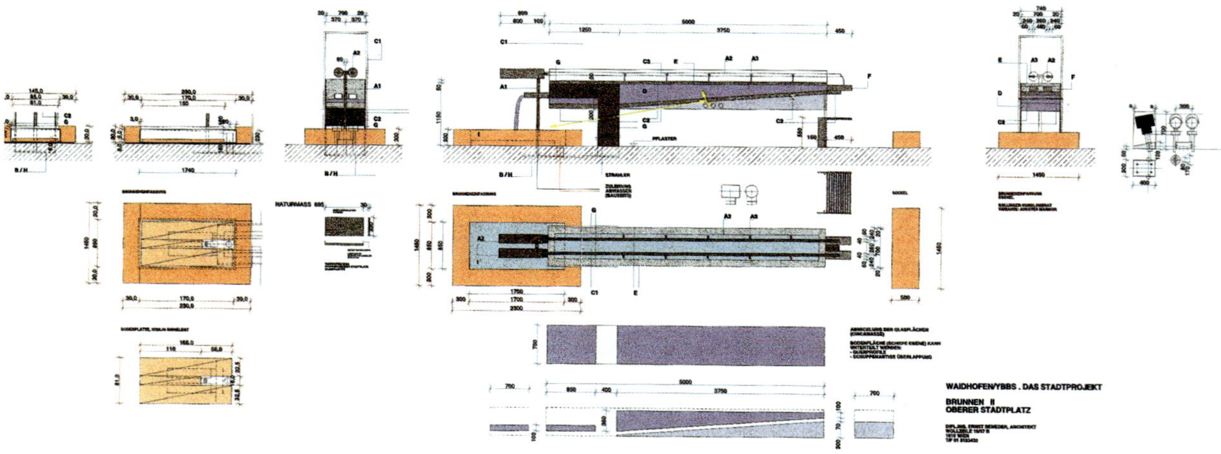

Neue Brunnen und Möblierungen
New fountains and furniture

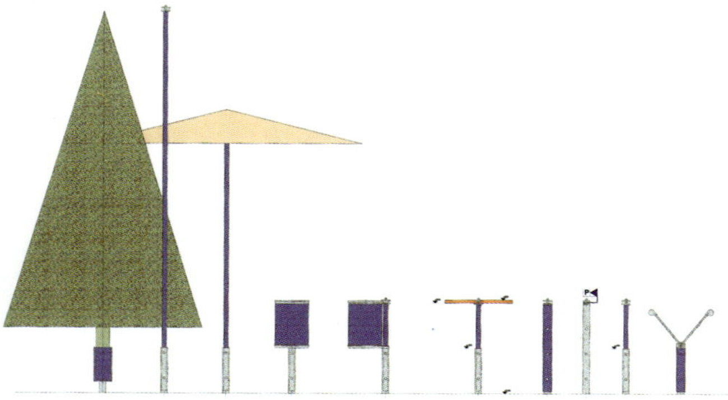

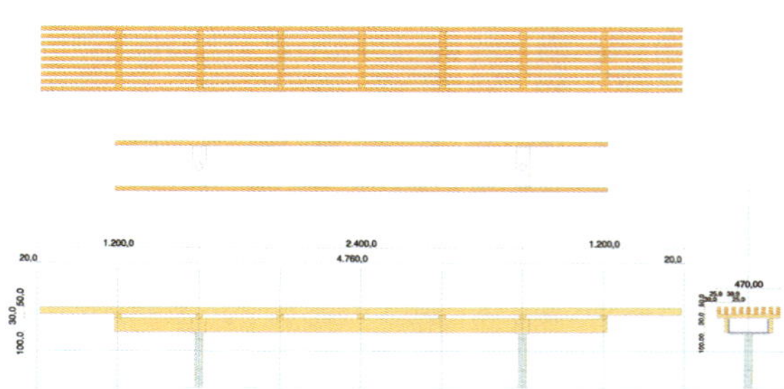

Rathaus

SORGNER

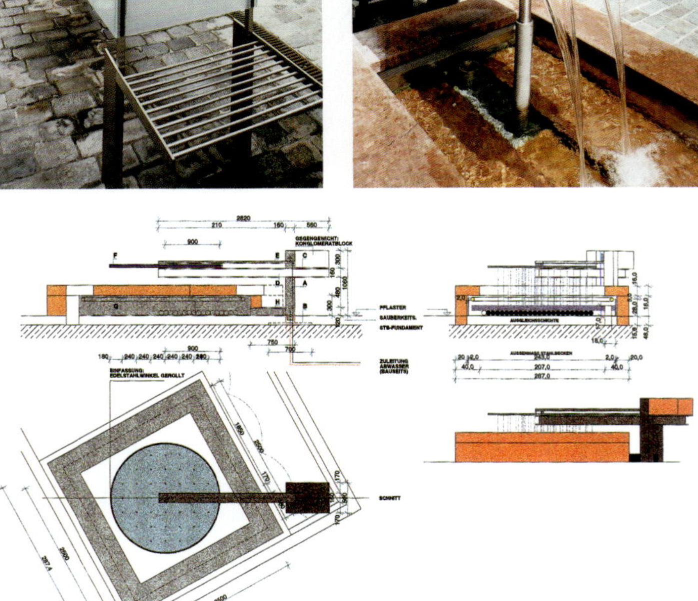

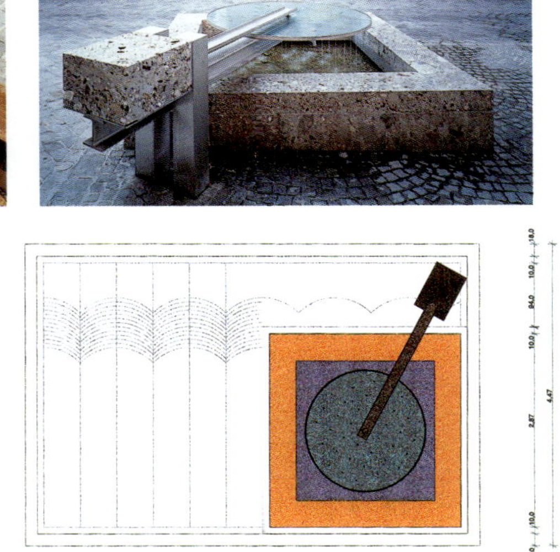

Der Weg ist das Ziel

Ein weiteres wesentliches Element des Erfolgs von Beneder war seine behutsame und jederzeit mehrheitsfähige Grunddefinition des Ziels des Stadtumbaus, nämlich das bewährte „Echte" zu erhalten: „Als ‚Echtes' sind für dieses Leitprojekt von Bedeutung: Räume (Talkessel, Platz- und Straßenräume, die Ybbsschlucht), die Nähe zur Natur, eine geglückte Maßstäblichkeit und die unaufdringliche Anmutung des Ganzen." Für die Innere Stadt wurde als Ziel angepeilt: „Wiedergewinnung der Stadträume als Lebensraum, öffentliche Nutzung vorhandener, das Stadtbild prägender Objekte sowie eine sozial ausgeglichene Bevölkerungsstruktur im Stadtzentrum." Nach der Erstellung des Leitprojekts (das auch die noch nicht realisierten Infrastrukturmaßnahmen einer Parkgarage und diverser Fußgänger-Querverbindungen der Stadträume durch Stege und Wege mit einschloss) wurden die geplanten Maßnahmen mittels Infotafeln am Ort der Intervention den Bürgern bekannt gemacht. Als erstes realisierte Beneder 1995 den Umbau der in einer Gruppe dichtgedrängter, kleiner, mittelalterlicher Häuser untergebrachten Stadtverwaltung zu einem „offenen" Rathaus. Die präzisen Interventionen und Sanierungseingriffe (*architektur.aktuell* 186) sollten nicht nur Service und Identifikation der Bürger mit ihrer Stadtverwaltung verbessern, sondern auch – was administrativ nur einer Statutarstadt möglich ist – eine Art „one-stop"-Erledigung aller „Amtswege" der Bürger bringen.

The Path Is The Goal

A further significant element of Beneder's success was his careful basic definition of the aims of urban renovation that always proved capable of attracting the support of the majority i.e. the preservation of the 'genuine' and historically tested. For the master plan the 'genuine' elements of significance include spaces (the floor of the valley, the squares and street spaces, the Ybbs ravine), the closeness to nature, a successful scale and the unimposing graciousness of the place as a whole. For the inner town the goal set was "the reacquisition of urban space as living space, the public use of existing objects that form the image of the town as well as a socially balanced population structure in the town centre". After the completion of the master plan (which also included as yet unrealised infrastructure measures such as a car park building and various pedestrian links through the town via footbridges and paths) the citizens were informed of the measures planned by means of information signs mounted at the site of the proposed interventions.

The first project to be carried out by Beneder in 1995 was a conversion job for the municipal authority that was housed in a group of small tightly planned mediaeval houses which he transformed into an 'open' Town Hall. The precise interventions and renovations works (*architektur.aktuell* 186) aimed not only at improving the level of service and the identification of the townspeople with the town administration but also at creating a kind of 'one stop' system where the townspeople can carry out all their dealings with the authorities in a single building, an idea of course possible only in a statutory town.

Beginn des Stadtprojekts (1995): Rathaus-Umbau
Start of the project (1995): town hall conversion

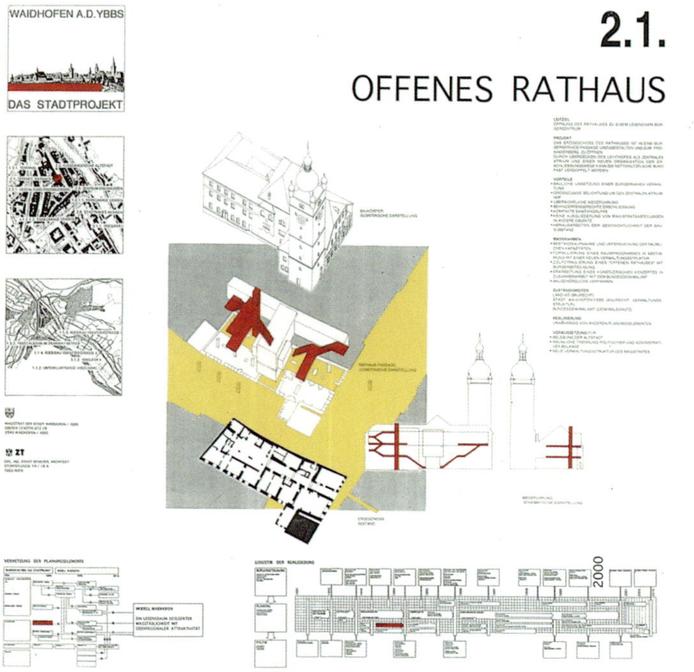

Nach Vorprojekten zur Umgestaltung der Inneren Stadt und einer Machbarkeitsstudie zur geplanten Parkgarage wurde als zweiter Schritt 1999 ein Infrastrukturprojekt realisiert, das einen schönen Natur-Erholungsraum für Einheimische und Besucher erschloss. Entlang des Ybbsufers, zwischen dem Wasserlauf und den steil abfallenden Felsen, auf denen die Stadt liegt, wurde ein Weg durch den Uferwald angelegt, der wegen der Felssturzgefahr teilweise überdacht werden musste. Diese Überdachung ist eine leichte, geräthafte Intervention des Designers Beneder, sie bietet mit einer Aussichtsplattform auch einen spektakulären Blick in das malerische Setting von Stadt und Naturraum.

Der Platz als Symbol

Als vorerst letzte Etappe des Stadtprojekts (Garage und Fußgängerstege fehlen noch) wurde nun die Gestaltung des Oberen und des Unteren Stadtplatzes fertiggestellt. Hier regte sich erstmals der Widerstand einiger Bürger in Form einer Volksbefragungsinitiative – weniger aus tatsächlichen funktionalen Gründen, da ja jedermann an der Stadtsanierung grundsätzlich interessiert war, sondern eher wegen diverser persönlicher Befindlichkeiten einflussreicher Bürger und Politiker. Waidhofen zeigte aber exemplarisch, wie auf eine solche Intervention zu reagieren ist. Die zuvor jahrelang betriebene intensive Kommunikationsarbeit für das Projekt trug Früchte in Form eines gemeinsamen Aufrufs aller Rathausfraktionen (man stelle sich vor: Selbst Freiheitliche und Grüne zogen an einem Strang), bei dieser Volksbefragung für das Projekt zu stimmen, was im März 2000 mit 78 % Zustimmung dann auch geschah und das Projekt letztlich unanfechtbar machte.
Beneder wandte in den Platzgestaltungen eine geradezu unscheinbare Gestaltungsphilosophie an, die vor allem auf die Rekonstruktion bestimmter Raumerlebnisse abzielte. Als typische Qualität der Stadt erkannte er den komplexen historischen Raum auf einem H-förmigen Grundriss, den der Obere und der Untere Stadtplatz – verbunden von einem Verbindungsplatz – gemeinsam bilden. Der Eingriff bestand letztlich „nur" in sanften Regulierungs- und Umgestaltungsmaßnahmen. Eine leichte Absenkung des Oberen Hauptplatzes vor dem mittelalterlichen Stadtturm – dem unbestrittenen Zentrum der Stadtkomposition –, subtile Pombierung der Oberfläche, Drainage, Verrohrung für bestehende und zukünftige Verkabelungen, eine neue „Wiener-Würfel"-Pflasterung, die differenziert auf räumliche Rhythmen reagiert, eine neue Platzbeleuchtung, die Einfügung von Halterungen für das temporär genutzte „Gerät" (Masten, Stände, Bänke, Bühnen u.a.), das gerade in einer ländlichen Kleinstadt bei den vielen identitätsstiftenden Märkten und Festen essentiell ist – und auch zwei Brunnen, die wieder alte öffentliche Platzqualitäten reaktivieren, die bereits in Vergessenheit geraten waren. Der Autoverkehr wurde reguliert, aber nicht gänzlich verbannt – was auch dem Stimulierungs- und Identifikationsabsichten gewissermaßen zuwidergelaufen wäre.

Following initial projects for the redesign of the town centre and a feasibility study on the planned car park building, the second step in 1999 was an infrastructure project that opened up a recreational area of natural beauty for both local residents and visitors. A path was laid out along the banks of the Ybbs, between the watercourse and the steep cliffs on which the town is built, through the riverside woods where it had to be roofed over in places to provide protection from falling rocks. This roof is a light, machine-like intervention by the designer Beneder. It has a viewing platform offering spectacular views of the picturesque setting of the town and the surrounding countryside.

The Square As A Symbol

The last stage of the town project up to the present (the garage and the pedestrian bridges are yet to be built) was the recently completed design of the Obere and Untere Stadtplätze (Upper and Lower Town Squares). Here for the first time opposition from the residents arose in the form of a referendum initiative that was due less to real functional considerations (as everybody was concerned that the town be renovated) but rather to the various personal sensibilities of influential townspeople and politicians. However in an exemplary manner Waidhofen demonstrated how to react to such an intervention. The intensive communication work which had been carried out over the years for the project bore fruit in the form of a combined call by all parties in the town hall (just imagine: even the Freedom Party and the Greens joined forces!) to vote in favour of the project in the referendum, which was carried with a majority of 78% in March 2000, a result that made the project unassailable.
For the design of the squares Beneder applied an unassuming design philosophy aimed primarily at the reconstruction of certain spatial experiences.
He acknowledged the complex historical space on an H-shaped plan formed by the Upper and Lower Main Squares (connected by a space in between) as one of the town's typical qualities. His intervention finally consisted 'only' of gentle regulatory and redesign measures. A slight lowering of the Upper Square in front of the mediaeval Stadtturm that is the uncontested centre of the urban composition, a subtle curving of the surface, drainage, underground conduits for existing and future cabling, 'Vienna Cube' paving that reacts in different ways to the spatial rhythms, new lighting, the introduction of a system of supports for those temporarily used 'appliances' (masts, stands, benches, stages etc.) – so essential in a small country town with all the local markets and festivals used to strengthen its identity – and two fountains that reactivated old qualities of the squares long forgotten. Vehicular traffic was regulated but not excluded completely as this would have worked to a degree against the intentions of stimulating and creating an identity.

Ybbsweg: Schutzüberdachungen und Infotafeln (1999)
Ybbsweg: protective roofing and information boards (1999)

Totale Identifikation

Da aber gerade der Obere Hauptplatz in hohem Maße Symbol für die Stadt als Ganzes ist, entzündeten sich erneut Widerstände an einem angeblich ästhetischen Problem, das hinter funktionalen Forderungen maskiert wurde. Beneders Brunnenentwurf (eine Glaswanne mit technoid gestylten Zu- und Ausflüssen, die ein wenig an Objekte von Günther Domenig erinnert und der Metallstadt Waidhofen materialsemantisch angepasst ist) wurde zum Schanigarten- und Verkehrshindernis stilisiert, obwohl an seinem subtil kalkulierten Standort vor dem Turm und in gemessener Distanz neben der barocken Mariensäule schon vorher „Gerät" stand (ein alter Wetterstand). Die geforderte Verschiebung wurde aber aufgegeben.

Das Stadtprojekt Waidhofen ist Resultat der glücklichen Konstellation von tradierten Bedingungen, einer emanzipierten Bürgerschaft und einem kommunikationsfreudigen Architekten mit idealem persönlichen Hintergrund. Die Stadtaufwertung durch eine Vielzahl kleiner Eingriffe, die in Summe große Wirkung erzielen, ist jedoch auch als Strategie übertragbar. Mit dem Gegebenen arbeiten, nichts von den vorhandenen positiven Qualitäten einer urbanen/ländlichen Situation verwerfen, in kleinen Schritten unter maximaler Kommunikation bedächtig voranschreiten, das ist die Strategie sowohl für Architekten als auch das „public management" der unmittelbaren Zukunft in Mitteleuropa. Bürgermeister Mair wagt sogar eine revolutionäre Einschätzung der dominierenden Wirtschaftsstruktur als bloße „Mode": Das Shopping in den weltweit gleichermaßen eintönigen suburbanen Malls werde vom anspruchsvollen zeitgenössischen Bürger dann aufgegeben, wenn es attraktive Angebote an traditionellen Standorten in authentischem Setting gäbe. Und noch eine Unglaublichkeit aus der Waidhofener Gegend: Ein Bauer verlangte die Rückwidmung eines Grundstücks von Bauland (Wert: 70 Euro/m^2) in Grünland (2 Euro/m^2), weil die Möglichkeit des Wirtschaftens mit eigenen Ressourcen ihm wichtiger erschien als der schnelle Profit. Bei dieser totalen Identifikation werden die umgebenden grünen Hügel, die das Stadtbild prägen, gewiss erhalten bleiben...

Total Identification

As to a considerable extent the Upper Main Square in particular is a symbol for the town as a whole, opposition again formed based on an allegedly aesthetic problem that in fact was used as a pretext for presenting functional demands. Beneder's design for a fountain (a glass basin with a technologically styled inflow and outlet, somewhat reminiscent of an object by Günther Domenig, in terms of materials semantically appropriate to the metal town Waidhofen) was made out to be a hindrance for outdoor cafes and a traffic obstacle. Although an 'appliance', a weather station, had previously stood at its subtly chosen location in front of the tower and at a measured distance to the Baroque Marian column. However the demand for the shifting of this fountain will be dropped.

The Waidhofen urban project is the result of a fortunate constellation of handed down constraints, an emancipated population and an architect with the ideal personal background and who enjoys communicating his ideas. The enhancement of a town by a series of smaller interventions that, when combined, achieve a major effect can also be applied as a strategy: working with the given situation, rejecting nothing of the existing positive qualities of an urban-rural situation, proceeding in small steps and employing the maximum of communication. This is a strategy for architects and 'public management' for the immediate future in Central Europe. The mayor Mr. Mair even ventured a revolutionary assessment of the dominant economic structure as mere "fashion": Shopping in suburban shopping malls, which are monotonously similar throughout the world, will be abandoned by critical contemporary townspeople as soon as attractive offers can be made at traditional locations in authentic settings. One further incredible fact from the area around Waidhofen: A farmer is currently requesting the rezoning of a site he owns from building land (value: 70 Euro/m^2) to green land (2 Euro/m^2) as the possibility of utilizing it with his own resources appears more important to him than making a quick profit. Given such total identification, the surrounding green hills that dominate the appearance of the area will certainly be preserved...

Technische Daten | technical data

Das Stadtprojekt. Gestaltung Innere Stadt
Waidhofen/Ybbs/Österreich, Oberer & Unterer Stadtplatz

Bauherr | client **Magistrat der Stadt Waidhofen/Ybbs**
Planung | planning **Ernst Beneder**
Örtl. Bauaufsicht | site supervision **Dipl. Ing. Heinz Plöderl**
Pflasterung | paving **STRABAG; AGes BAU**
Asphalt **Swietelsky**
Elektroinstallationen | electrical services **Rigler Elektrik**
Metallbau | metal construction **Metallbau Seyrlehner; Metallbau Mittermair**
Lichtplanung/-ausstattung | lighting concept/fittings **Lighting Design Austria**

Grundstücksfläche | site area	15.000 m^2
Nutzfläche | floor area	15.000 m^2
Wettbewerb | competition	1991
Leitprojekt | masterplan	1992
Planungsbeginn | start of planning	1999
Baubeginn | start of construction	4/2000
Fertigstellung | completion	11/2000
Brunnen | fountain	5/2001
Baukosten | building costs	ca. 4,36 Mio EUR
Kosten pro m^2 | cost per m^2	ca. 290,- EUR

Dominique Perrault

Dominique Perrault/RPM

Rathausgalerien in Innsbruck, Austria

Photos Roland Halbe
Text Matthias Boeckl

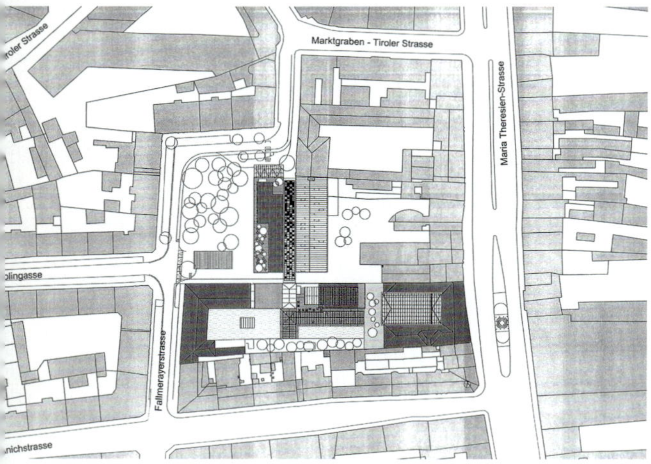

Hybrider Elementarismus

Intensive Suburbanisierung, Reduktion der historischen Kernstadt auf eine Tourismuskulisse einerseits und öffentliche Funktionen als „Hebel" ihrer Sanierung durch Einführung alt-neuer kommerzieller Funktionen andererseits – das war das Szenario beim aktuellen Innsbrucker Stadtumbau. Die Rathausgalerien realisieren das derzeitige Bild von „Öffentlichkeit": Konsum und Entertainment als einzige Möglichkeit, dafür heute noch eine Basis zu generieren.

Hybrid Elementarism

Intensive suburbanisation and the reduction of the historic core to a tourist backdrop on the one hand and public functions as a quasi-lever for refurbishment through the introduction of old-new commercial functions on the other. This was the scenario in a recent Innsbruck urban revitalisation project. The Rathausgalerien (Town Hall Galleries) are the manifestation of a contemporary image of the 'public': consumption and entertainment as the sole possibility of generating a basis for this quality today.

Das Projekt der Rathausgalerien ist ein perfektes Beispiel für den heute notwendigen Funktions- und Investorenmix in der Stadtsanierung. Diese hybride Struktur ist ein Abbild unserer vielfältigen Gesellschaft, in der nicht mehr nur ein einziges kulturelles Bild das Zusammenleben dominieren kann. Die Altstadt von Innsbruck ist wie jede andere historische Kernstadt einem massiven Transformationsprozess unterworfen. Denn auch in Innsbruck hat sich ein Ring von Einkaufszentren um die Stadt gelegt und auch in Innsbruck wollen jene, die es sich leisten können, nicht mehr unbedingt in der von Parkplatznot und anderen Problemen belasteten Altstadt wohnen. Die Folgen sind bekannt und können nur mehr – wie es auch das Beispiel von München und Berlin zeigt – mit einer teilweisen Preisgabe der kompromisslosen Konservierung eines historischen Stadtbildes und mit der Schaffung neuer kommerzieller Attraktionen gemildert werden. Innsbrucks Innenstadt ist zusätzlich noch mit der speziellen österreichischen Situation des intensiven Tourismus konfrontiert, der hier konkret aus Kurzbesuchen des historischen Quartiers um Hofburg und Goldenes Dachl mit den bekannten Folgen für Wirtschaftsstruktur und Population besteht.

The Rathausgalerien project is a perfect example of the mix of functions and investors required for urban refurbishment. Such a hybrid structure is a reflection of our diverse society in which a single cultural image of how we live together can no longer dominate. Like all other historic cores the old town in Innsbruck has been subjected to a massive transformation process. In Innsbruck too a ring of shopping centres has been formed around the town, and those who can afford other options do not always want to live in the old town with its lack of car parking spaces and other problems. The consequences are well known and – as the examples of Berlin and Munich demonstrate – can only be alleviated by partially giving up the uncompromising conservation of the city's historic image and by creating new commercial attractions. In addition, Innsbruck's old town is confronted with the particularly Austrian phenomenon of concentrated tourism, which in concrete terms consists here of brief visits to the historic areas around the Hofburg and the Goldenes Dachl with the all too familiar implications for the economy and the population.

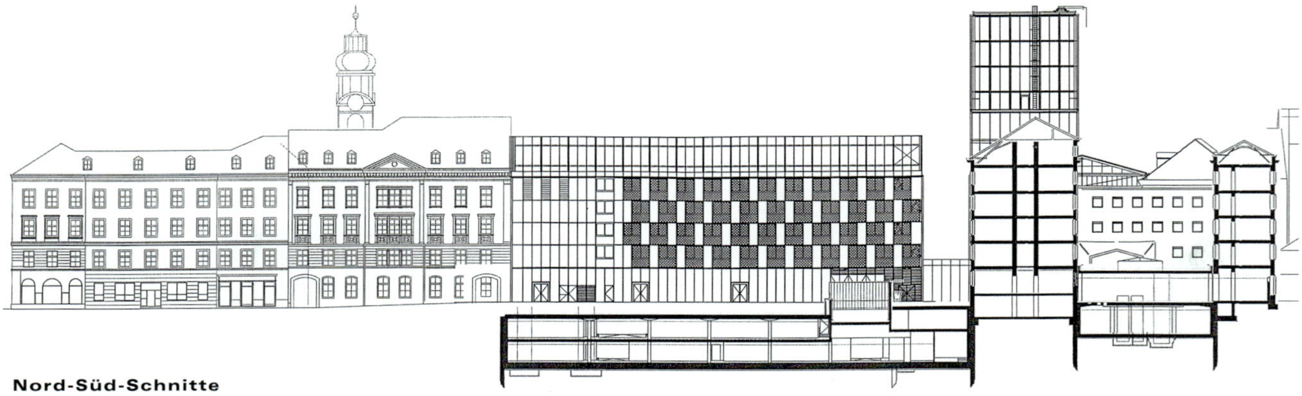

Nord-Süd-Schnitte
Sections, north-south

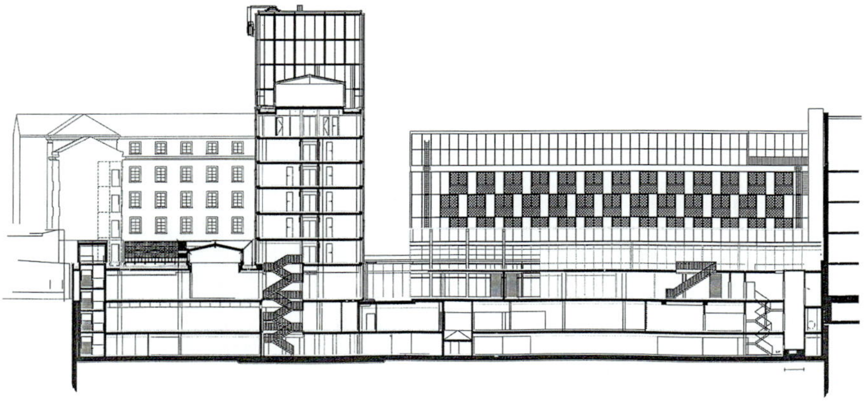

Nordansicht des Komplexes
North view of the complex

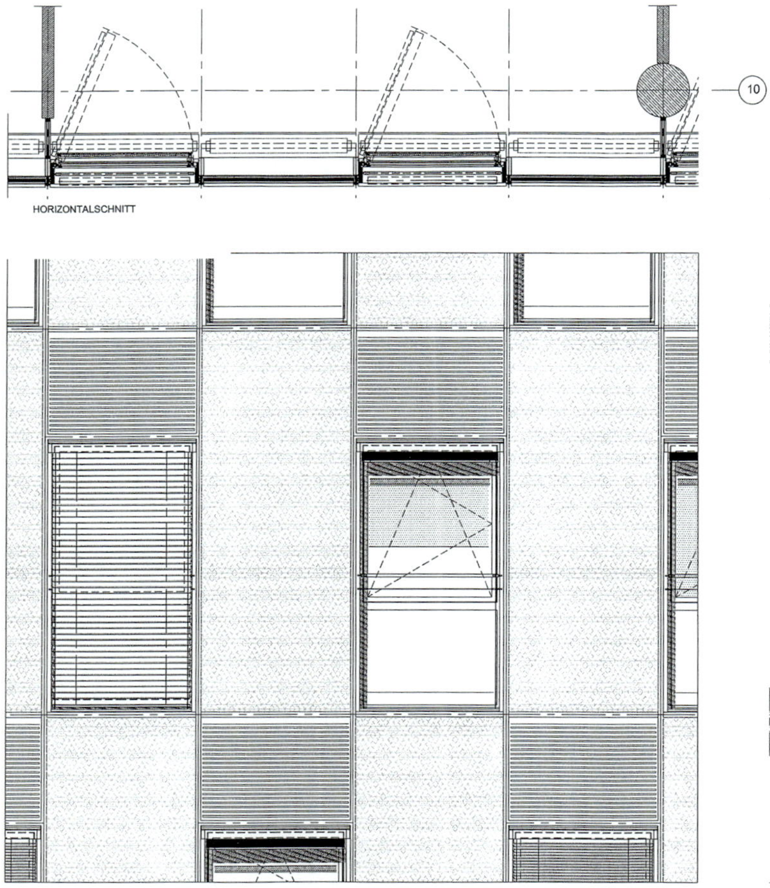

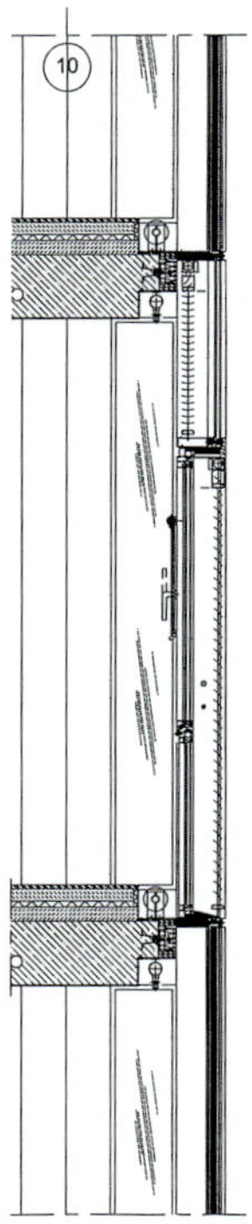

Fassaden der Bürogeschoße; Dachlandschaft der Galerien
Facades of the office levels; roofscape of the galleries

Elementarismus als Lösung

Mit dieser Rahmensituation war die Stadtverwaltung gleichsam auch am „eigenen Leib" konfrontiert, da der zentral gelegene, aus mehreren Gebäuden bestehende Rathauskomplex durch Kriegsschäden, chronische Platznot und Abnutzung seit langem eine offene städtebauliche Wunde der Stadt war. Man versuchte schon 1985 in einem Wettbewerb dafür Lösungen im Sinne einer Gesamtsanierung zu finden, aber erst ein zweiter Wettbewerb im Jahre 1995 brachte mit Dominique Perraults Projekt eine Realisierungsperspektive. Die übrigen Teilnehmer waren Guido Canall (I), Günther Domenig (A), Galfetti-Könz (CH), Massimiliano Fuksas (I), Josef Lackner (A), Peter Lorenz (A), Rüdiger Kramm (D) und Heinz-Mathoi-Strehli (A).

Perrault musste wegen der komplexen Anforderungen an das Projekt, in dem mehrere private und öffentliche Investoren gemeinsam einen gemischt genutzten, innerstädtischen Komplex, teilweise in einer Hinterhofsituation mit großen Anteilen renovierter Bestandsbauten, aber auch mit einer Neudefinition des Adolf-Pichler-Platzes errichteten, ungewohnt auf eine Innenwirkung seiner bekannt geometrisierenden Formensprache setzen. Nicht zuletzt deshalb werden viele interne Blickachsen durch Glasdecken inszeniert.

Überraschend ist es aber, dass der von starker künstlerischer Suggestivkraft getragene Elementarismus von Perrault, der noch dazu stets in „High-Tech"-Gewand auftritt, in einem kleinteiligen Altstadtgewebe keineswegs megaloman oder unangebracht wirkt, sondern im Gegenteil die lang erhoffte Klärung und Sanierung der Situation bei großer öffentlicher Akzeptanz gebracht hat.

Elementarism As A Solution

The municipal administration was confronted with this situation, at its own front door so to speak, as the centrally located Town Hall complex, made up of several buildings, had some time ago become an open urban wound as the result of war damage, a chronic shortage of space and general wear and tear. In 1985 an attempt had been made to find a solution with a competition for a general refurbishment but it was only the second competition, held in 1995, that offered perspectives of realising this goal in the form of Dominique Perrault's entry. The other participants in the competition were Guido Canall (I), Günther Domenig (A), Galfetti-Könz (CH), Massimiliano Fuksas (I), Josef Lackner (A), Peter Lorenz (A), Rüdiger Kramm (D) and Heinz-Mathoi-Strehli (A). Given the complicated requirements of this project, in which several private and public investors together erected a mixed use inner city complex, partly in a backyard situation with a large proportion of renovated existing buildings but also involving a redefinition of Adolf-Pichler Platz, Perrault had, unaccustomedly, to rely on the internal effects of his familiar geometric formal language. Not least for this reason a number of internal visual axes were created by the use of glass ceilings. Yet it is surprising that Perrault's elementarism, which possesses a strong artistic suggestive strength and furthermore is always clad in 'high-tech' garb appears neither megalomaniac nor inappropriate in the small-scale mesh of the old town but, on the contrary, has achieved the long awaited clarification and improvement of the situation and gained considerable public acceptance.

Sonnenschutz aus Edelstahlgewebe
Sun protection, stainless steel mesh

Dachterrasse des Hotels
Hotel roof terrace

1 **Hotel**
2 **Aufgang zu den Rathausbüros**
Access to the town hall offices
3 **Supermarkt |** supermarket
4 **Rathaus – Bestandstrakt (1938)**
Town hall, existing tract
5 **Altes Rathaus |** old town hall
6 **Geplante Erweiterung zur Anichstraße**
Planned extension to Anichstrasse
7 **Shopping Mall**
8 **Neugestalteter Platz mit Tiefgarage**
Redesigned square with underground carpark

Hotel und Nordzugang | hotel and north approach

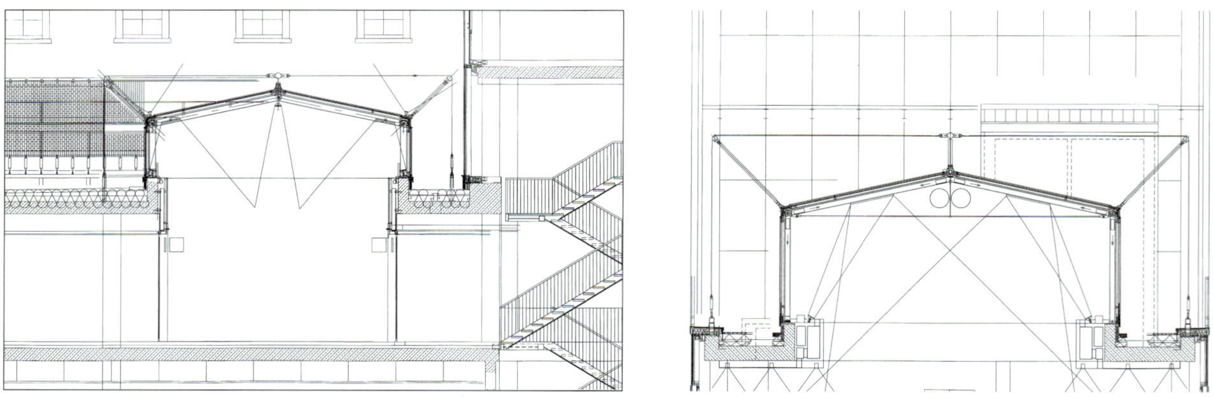

Überdachung der Galerien | roofing of the galleries

Wegkreuzung | crossing of Galleries

Innenhof beim Alten Rathaus | courtyard at the old town hall

Jones
RLB

Dieser Erfolg war aber kann denkbar ohne eine starke ökonomische Grundstruktur, deren Koordination vom einzigen österreichischen „Corporate Office" – also einem großen Gesamtanbieter von Planungsleistungen mit angeschlossenem Immobiliengeschäft – betreut wurde, und zwar von der Planungsfirma Achammer, Tritthart & Partner (ATP). Erst in der Partnerschaft mit Reichert-Pranschke-Maluche aus München und ATP aus Innsbruck konnte Perraults Wettbewerbsprojekt zur Baureife gebracht werden. Die Grundstruktur besteht – vereinfacht gesagt – darin, dass die Stadtverwaltung in einer „Private-Public Partnership" den Baugrund einem Investorenkonsortium aus Banken und Geschäftsleuten zur Verfügung stellt, das darauf eine Erweiterung und Sanierung des weitgehend unbrauchbar gewordenen alten Rathauskomplexes, eine Shopping Mall und ein Hotel errichtet. 6.000 Quadratmeter Rathausbüros wurden dann der Stadt zurückvermietet.

Markt als Urbild der urbanen Öffentlichkeit

Die vielfältigen Funktionen mussten im Inneren eines großen urbanen Blocks untergebracht werden, der die Anlage auf drei Seiten umschließt. An der offenen vierten Seite, die an einen bislang eher von Randgruppen genutzten Platz grenzt, wurde das Hotel mit 96 Zimmern positioniert. Ein Penthouse mit Restaurant bietet hier höchststehende touristische und gastronomische Wonnen. Zwischen Hotel und den drei anderen Innenseiten des Karrees platzierten Perrault und Partner um und über die kreuzförmig in das Karree eingeschriebenen Gänge 22 Geschäftslokale, mehrere Gastronomiebetriebe und die Rathausbüros. Am Schnittpunkt der beiden Wegachsen, die auch als innerstädtische Verbindungen der besseren „Durchlüftung" der Stadt dienen und wie die alten Galerien transparent überdacht sind, erhebt sich der markante, 38 Meter hohe Glasturm des Rathauses. Unmittelbar daneben liegt – unscheinbar zwischen den Geschäftslokalen verborgen – der architektonisch eher unterbewertete Haupteingang zum Rathaus über einem mehrfach umbrochenen Treppenlauf.

This success was, of course, unimaginable without a strong economic basic structure that was coordinated by Austria's only 'corporate office', i.e. major supplier of a package of planning services with an associated real estate company, the planning company Achammer, Tritthart & Partner (ATP). It was only in partnership with Reichert-Pranschke-Maluche from Munich and with ATP from Innsbruck that Perrault's competition project could be developed to the building stage. Put simply, the basic structure was as follows: In the context of a private-public partnership the municipal administration made the development site available to a consortium of investors composed of banks and business people, who then carried out an extension to and renovation of the old and, for the most part, unusable Town Hall complex as well as erecting a shopping centre and a hotel; 6,000 square metres of Town Hall office space was then rented back to the town.

The Market As A Primal Image Of Urban Public Life.

These diverse functions all had to be housed within a large urban block that encloses the complex on three sides. A hotel with 96 bedrooms was placed on the open fourth side, which adjoins a square that was generally frequented by the urban subculture until recently. A penthouse with restaurant offers the height of gastronomic and other tourist delights. Between the hotel and the other three internal sides of the block Perrault and Partners positioned 22 shops, several restaurants and the Town hall offices around and above the passageways inscribed in the site in the shape of a cross. The striking 38 metre high Town Hall glass tower rises at the intersection of the two axes – which also serve as inner city connections providing better 'ventilation' for the town. Modestly hidden amongst the shops and directly beside it is the architecturally too understated main entrance to the Town Hall via a segmented staircase that changes direction several times.

Rathausturm mit Glasgestaltung von Peter Kogler
Town hall tower with glass design by Peter Kogler

Diese in den Obergeschoßen situierten Büros verbinden in einer langen Achse zwei gegenüberliegende Altbauten des Karrees, die schon bisher der Stadtverwaltung gedient hatten und teilweise unter Denkmalschutz standen (eines davon stammt aus der NS-Zeit).

Kunst-Fassaden und -Dächer

Die verbindenden neuen Baukörper Perraults sind in einer linearen geometrischen Linienführung und farblich sowie materialmäßig in einem technoiden Stil gehalten. Es dominiert der Schwarz-Weiß-Akkord, der in der Färbelung der sanierten Altbauten weitergeführt wird. Die Haupteffekte der Neubaufassaden verdanken sich aber einem als Sicht- und Sonnenschutz genutzten Stahlgewebe und gläsernen Aufbauten, die als Restaurant und Aussichtsplattform genutzt werden. Durch den gesamten Komplex ziehen sich künstlerische Interventionen, die von Silvia Eiblmayr, der Leiterin der zeitgenössischen Landesgalerie, und dem Direktor des Wiener Museums moderner Kunst, Edelbert Köb, kuratiert wurden. Am intensivsten wirken hier die mit dem Tageslicht sich verändernden färbigen Glasfelder in einer Hofüberdachung von Daniel Buren sowie die mit Motiven von Peter Kogler bedruckten Glaswände des zentral situierten Rathausturms. Im neuen, eleganten Gemeinderatssitzungssaal, der im Dachgeschoß untergebracht ist, von wo aus die Stadtväter ihren Wirkungskreis überblicken können, betont eine Schriftinstallation von Heinz Gappmayr die hier verhandelte Balance zwischen Gemeinschaft und Individuum.

On a long axis, these offices located on the upper floors connect two old buildings lying opposite each other that were previously also used by the municipal administration and are partly under a preservation order (one of these dates from the Nazi period).

Art Facades And Roofs

Perrault's connecting elements are linear geometric forms, in terms of both colour and material essentially technoid in style The harmony between black and white dominates and is continued in the colouring of the refurbished old buildings. The main effect of the new building facades is achieved by the use of a steel mesh, employed as visual screening and solar protection, and glass rooftop elements that are used as a restaurant and viewing platform. The artistic interventions to be found throughout the entire complex were curated by Silvia Eiblmayr, the director of the Innsbruck Gallery for Contemporary Art and by Edelbert Köb, Director of the Museum of Modern Art in Vienna. Among the most intensive of these are the coloured glass bays by Daniel Buren, positioned in a glass-roofed courtyard, that alter their appearance according to the changing daylight and the glass walls of the centrally situated Town Hall tower printed with a motif by Peter Kogler. In the elegant new council meeting room located at roof level, from where the city fathers have an overview of their field of activities, an inscription installation by Heinz Gappmayr emphasises the balance between the community and the individual that has been negotiated here.

Technische Daten | technical data

Rathausgalerie
Innsbruck/Österreich, Maria Theresien Straße 18

Bauherr | client **Rathauspassage GmbH**
Projektentwicklung/-management | project development & management **BOE Bauobjekt Entwicklung GmbH & Co KG**
Architektur-Planung | planning **Dominique Perrault & Reichert-Pranschke-Maluche**
Projektleitung | project supervision **Ralf Levedag; Alexander Stecher**
Mitarbeiter | assistance **Anne Speicher, Jerôme Thibault, Mathias Fritsch, Moreno Maconi, Cyril Lancelin; Nina Höhne, Rainer Centmayer, Pedro Castro-Ricardo, Stefan Hub, Thomas Bengesser**
Künstlerische Direktion | art director **Gaelle Lauriot-Prèvost**
Gesamtplanung | general concept **ATP – Achammer, Tritthart & Partner**
Statik | structural consultants **Karl Frajo-Apor, Stefan Demetz, Milan Kozanek, Manfred Lang, Margarita Pavlova, Heinz Steffan, Sabine Tussetschläger**
Kunst am Bau/Kuratoren | curators, artworks **Dr. Sylvia Eiblmayer, Prof. Edelbert Köb**
Künstler | artists **Daniel Buren, Peter Kogler, Prof. Heinz Gappmayr, Isa Genzken, Josef Dabernig, Ricarda Denzer, Martin Gostner, Dorit Margreiter, Matt Mullican, Prof. Walter Obholzer, Prof. Eva Schlegel, Milica Tomic-Stojanovic, Ernst Tragöwer**
Landschaftsarchitekt | landscape architect **Landscaping Ludwigstorff & Hösel**
Edelstahlgewebe | stainless steel mesh **GKD, Düren**
Lichtplanung | lighting concept **Bartenbach Lichtlabor GmbH, Aldrans**
Lichtausstattung | light fittings **Durlum, Innsbruck**

Grundstücksfläche | site area	12.960 m²
Nutzfläche | floor area	35.200 m²
Bebaute Fläche | built-up area	48.000 m²
Umbauter Raum | cubage	180.800 m³
Wettbewerb | competition	1996
Planungsbeginn | start of planning	1997
Baubeginn | start of construction	2000
Fertigstellung | completion	2002

Graz 2003

Graz 2003

Städtebau als Kommunikations-Kunst
Urban Development As An Art Of Communication

Text Matthias Boeckl
Photos Angelo Kaunat, Paul Ott

Vito Acconci: Mur-Insel; Cook/Fournier: Kunsthaus

Ist es noch Illusion oder schon Realität des postindustriellen Medienzeitalters, dass ein grundlegender und nachhaltiger Stadtumbau eher durch „weiche“ Kulturfunktionen als durch traditionelle „harte“ Jobfaktoren wie Industrie und Gewerbe gelingen kann? Und was wäre dabei unter „Kultur“ zu verstehen? Eine neue Software für die Hardware Stadt? Welche Balance besteht zwischen temporären Festival-Interventionen und langfristigen Auswirkungen solcher Veranstaltungen? Handelt es sich um eine der berüchtigten Sternschnuppen der Festivalisierung der Städteplanung oder um ein tatsächlich neues Profil, das sich eine Stadt verschreibt?

Is it still an illusion or is it already the reality of a post-industrial media age that fundamental and lasting urban redevelopment can be achieved more through 'soft' cultural functions than the traditional 'hard' job factors such as industry and trade? And what is to be understood under 'culture'? A new software for the hardware city? What sort of balance exists between temporary festival interventions and the long-term effects of such events? Is this simply one of the notorious shooting stars of making a festival of urban development or a genuine new image to which the town is committing itself?

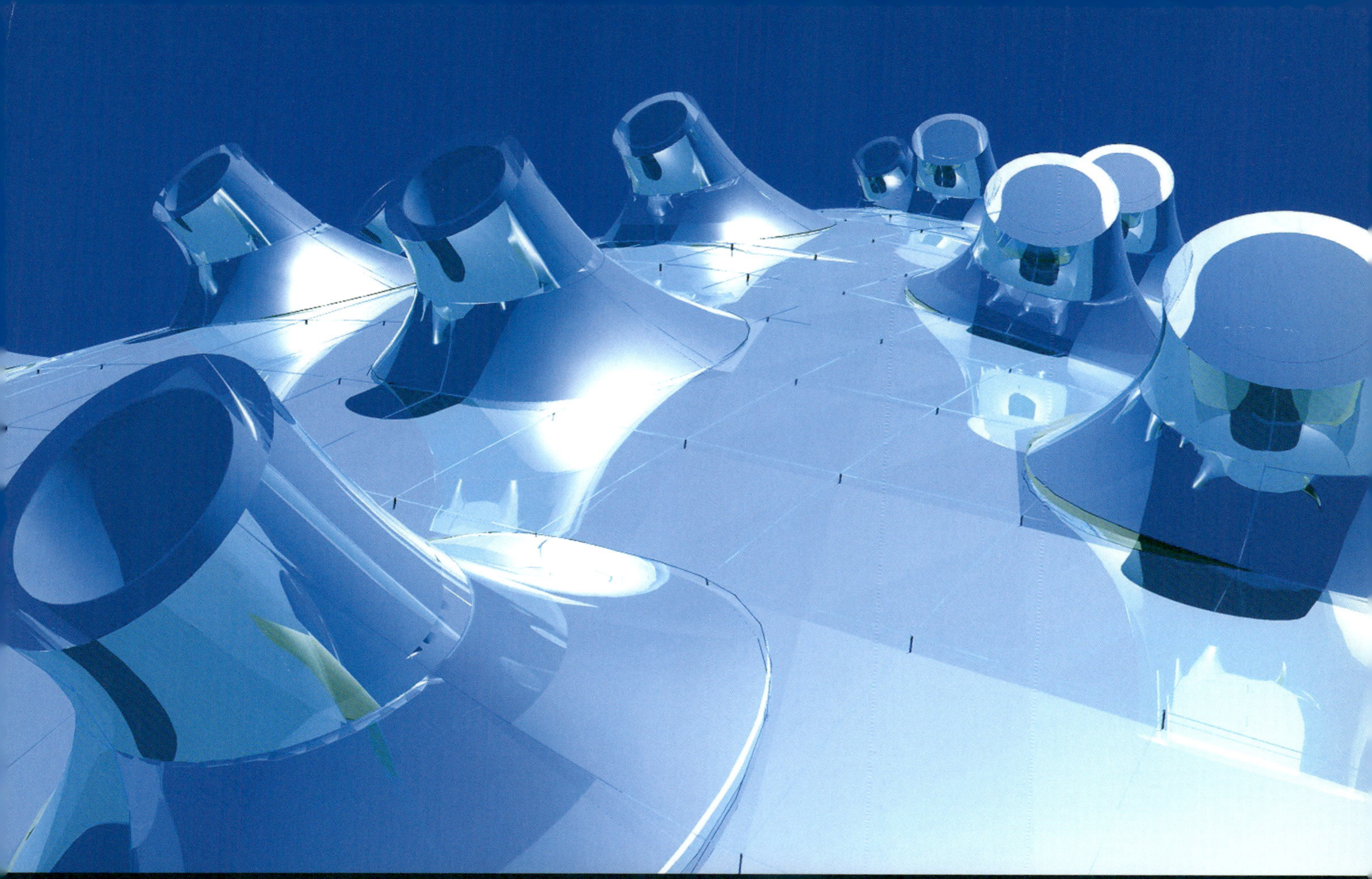

Die „europäische Kulturhauptstadt Graz 2003" bietet viele Antworten auf diese brennenden Fragen der aktuellen Stadtdebatte. Sie versteht sich, worauf Wolfgang Lorenz, Fernsehmacher und Programmintendant der Hauptstadt-Veranstaltungen, insistiert, eben nicht als „Festival", sondern als eine Art Bewegung zur Formulierung einer neuen Stadtidentität, eines neuen Selbstverständnisses der Grazer. Der Anlass für die zahlreichen Maßnahmen im Bereich Architektur und öffentlicher Raum war die Ernennung zur (einzigen) „Kulturhauptstadt" Europas für das Jahr 2003. Was das bedeutet, weiß niemand, denn mit der Ernennung ist keine feste Programmplanung verbunden. Jede Stadt muss ihre Inhalte selbst definieren. Und meist tut sie das in Form eben eines Festivals, das interessante Produktionen aller Art einkauft und ein Jahr lang auf bestehenden Bühnen abspielt. Jede Stadt hat aber auch andere Problemlagen, manche Städte haben intakte urbane Strukturen, funktionierende öffentliche Institutionen und Infrastruktur, ausreichend Arbeitsplätze, Wohnungen und Freizeiteinrichtungen, viele jedoch nicht.

Zwischen bürgerlicher Idylle und Avantgarde

Was waren die „Probleme" von Graz vor der „2003"-Bewegung? Die Landeshauptstadt der Steiermark hatte unter den europäischen Städten mittlerer Größe (226.244 Einwohner) schon bisher eine eigenartig ambivalente Positionierung. Graz, das stand lange Zeit für eine gewisse Isolierung in der südöstlichen Ecke des „freien" Europa – ein Image, gegen das man schon seit den 1960er Jahren mit einem emsigen, landesoffiziell gesponserten Kulturbetrieb ankämpfte, der über die Grenzen zu Ungarn und Slowenien hinweg unter dem Titel „trigon" bekannt war. Graz lebte stets aus ziemlich eindeutigen Gegensätzen. Hier die beeindruckende Altstadt und die konservative Bourgeoisie nebst untergründigen NS-Nostalgien, da die „Avantgarde", die es leicht hatte, mit Realismus und Aktionismus jede bürgerliche Idylle zu verstören. In diesem Klima entstand die Grazer Autorenversammlung (ein Flaggschiff der jüngeren deutschsprachigen Literatur), die „Grazer Schule" der Architektur mit ihren Pionieren Günther Domenig, Eilfried Huth, Szyszkowitz-Kowalski und anderen, sowie eine Vielzahl anderer künstlerischer Revolten, auch in Malerei und Plastik.

The European Cultural Capital Graz 2003 offers many answers to these burning questions of topical urban discussion. As Wolfgang Lorenz, TV producer and programme director of the regional capital's events, emphasises, it does not intend to be a festival at all but a kind of movement towards the formulation of a new urban identity, a new consciousness for the people of Graz. The numerous measures in the field of architecture and public space were occasioned by Graz being designated the (only) cultural capital in Europe for 2003. Nobody knows what this actually means as there is no fixed programme plan connected with the designation. Every town has to define the contents itself. As it happens, this is mostly done in the form of a festival which buys in interesting productions of every kind and runs them over existing stages for the duration of a year. Every town has different problem areas, though some cities have intact urban structures, functioning public institutions and infrastructure, enough jobs, flats and leisure facilities, but many do not.

Between Bourgeois Idyll And The Avant-Garde

What were the 'problems' Graz faced before the '2003 movement'? Until now the federal provincial capital of Styria has occupied a singularly ambivalent position among medium-sized European cities (226,244 inhabitants). For a long time Graz stood for a certain isolation in the most south-eastern corner of free Europe, an image which has been countered with a bustling culture industry under official provincial sponsorship since the 1960s, extending across the Hungarian and Slovenian borders and known under the title 'Trigon'. Graz always lived from quite definite opposites. Here the impressive old town and the archly conservative bourgeoisie existed alongside underground Nazi nostalgia as the avant-garde, which had it easy, destroyed this bourgeois idyll with realism and actionism. In this climate the Graz assembly of authors came into being (a flagship of young German literature), as did the Grazer Schule of architecture, with its pioneers Günther Domenig, Eilfried Huth, and Szyszkowitz-Kowalski among others, as well as a number of other artistic revolts, including those in painting and sculpture.

Peter Kogler:
Bahnhofs- | railway station **Mural**

Klaus Kada:
Stadthalle | convention Center

EINGANG
messecentergraz

Das war der Kunstsektor – und wirtschaftlich stand die Stadt stets unter dem starken Einfluss ihrer großen Industriebetriebe und ihrer natürlichen Funktion als administratives Regionalzentrum. Beides wirkte sich eher konservierend aus, bis in die 1980er Jahre waren hier kaum Faktoren wirksam, die zu einer regelrechten Stadttransformation hätten führen können. Nur die europaweit übliche schleichende Entvölkerung der Kernstadt durch Suburbanisierung schuf einen gewissen Handlungsbedarf: 1970 bis 1990 schrumpfte die Stadtbevölkerung um ca. 2,3 Prozent pro Dekade, 1991-2000 steigerte sich das Entvölkerungstempo auf fast fünf Prozent. Die Ostöffnung hatte also zunächst weder bevölkerungsmäßig noch im Kulturbetrieb sichtbare Spuren hinterlassen, was sich nun wohl ändern wird.

Kultur: urbaner Kick-Off oder bloß Kulisse?

Mit den Projekten zu „Graz 2003" könnte der Kick-Off zu einer wirklich zeitrichtigen Stadtentwicklung entstanden sein. Und hier hat man die Chance, bereits aus den Gefahren des „New Economy"-Hype zu lernen. Keine Kultur ausschließlicher Virtualität künstlich hochzufahren, sondern das Kunststück zustande zu bringen, physisch greifbare Stadtinterventionen in Form von Bauten und Infrastruktur mit der „virtuellen" Stadt einerseits und dem gewachsenen Bestand andererseits derart zu verweben, dass (etwa durch Kulturtourismus) sowohl rasche Aufwertungseffekte entstehen als auch Keime für nachhaltige Entwicklungen und die Möglichkeit der Änderung und Anpassung der nun entworfenen Strukturen.

This was the art sector, and the city remained under the strong influence of its large industries and its natural function as the regional administrative centre. Both had a relatively conservational effect well into the 1980s as there were few factors at play here which could have lead to a proper transformation of the city. Only the gradual depopulation of the city centre through sub-urbanisation, common throughout Europe, created any need to take action: from 1970 to 1990 the city's population shrank by approximately 2.3 per cent each decade, and between 1991-2000 the rate of depopulation rose to almost 5 per cent. At first the opening of Eastern Europe had no apparent effect on the population and culture industry, something which will now change.

Culture: An Urban Kick-Off Or A Mere Stage Set?

The Graz 2003 projects prompted the kick-off for genuine urban development in keeping with the times. This provides a chance to learn from the dangers of New Economy hype. Not to artificially promote a culture of exclusive virtuality but rather to actually see the artwork through, to interweave physically tangible urban interventions in the form of buildings and infrastructure with the virtual city on the one hand and the existing town structures on the other in such a way that (for example, through cultural tourism) rapid revaluation effects are created as well as seeds sown for lasting developments and the possibility for change and adaptation of the structures presently being designed.

Markus Penthaler: Helmut-List-Halle

Foyer

Das klingt nach der Quadratur des Kreises, in der Praxis zeigt sich aber, dass Stadtumbau durch Kulturfunktionen tatsächlich die eminente Gefahr der Reduktion einer historischen Kernstadt auf eine nur saisonal genutzte und von langjährigen Einwohnern klinisch gesäuberte Kulisse für den Tagesausflug der kulturinteressierten A-Schicht oder den einfachen Bummel anderer Besucher wird. Diese Gefahr ist in Innsbruck und Salzburg bereits zur Realität der Verödung der historischen Kernstadt geworden, wobei im Falle Innsbrucks die Stadtverwaltung durch alltägliche Funktionen, die sie hier neu ansiedelt (Verwaltung, Shopping), gegenzusteuern sucht, Salzburg jedoch eher auf die Ausweitung des elitären Kulturprogramms zu den Interessen der Lokalbevölkerung denn auf eine profunde Revitalisierung der Altstadt setzt.

„Dinge, die die Stadt braucht, aber auch Dinge, die purer Luxus sind."

„Graz 2003" brachte weniger durch einen langjährig von Stadtplanern erarbeiteten Masterplan die (hoffentlich funktionierende) Wende zustande, sondern eher durch den Termindruck, der die Stadtverwaltung förmlich zwang, den Ideen von Programmintendant Lorenz zu folgen, um nicht zum Eröffnungszeitpunkt ohne spektakuläre Events und Institutionen dazustehen. Diese Zwangslage verschaffte dem Intendanten großen Einfluss, brachte aber die Stadtpolitiker auch zu positivem Umdenken und letztlich zur Identifikation mit den Projekten. Neben zahlreichen Kunst-, Literatur- und Theaterveranstaltungen gibt es gerade auf dem Sektor Architektur und öffentlicher Raum ein richtiggehendes Feuerwerk an Projekten, die ohne Hauptstadt-Erfolgsdruck im stadtpolitischen Alltag niemals realisierbar gewesen wären. Graz genehmigte sich nun mit zunehmender Lust an dieser Entwicklung „Dinge, die die Stadt braucht, aber auch Dinge, die purer Luxus sind" (Lorenz).
An Bauten entstanden sowohl Projekte, die direkt auf Initiative der Programmplanung zurückgehen als auch solche, die gleichsam parallel dazu mit Rückenwind des zunehmend dynamischen Planungsgeschehens von unabhängigen Bauträgern errichtet wurden. Das Bauspektrum umfasst das Litera-

This sounds like the squaring of the circle, but in practice it is clear that urban development through cultural factors actually present the eminent danger of reducing a historic town centre to a stage set, used only seasonally and cleansed of any permanent residents, there merely for the day trips of the better class of the culturally interested or for other visitors to stroll around. In Innsbruck and Salzburg this danger has already become the reality of desolation in the historic centre. In the case of Innsbruck, the city administration is attempting to counteract this by relocating everyday functions there (administration, shopping), while Salzburg relies more on extending the cultural programme to embrace the interests of the local population rather than on a profound revitalisation of the old town.

"Things The City Needs But Also Things That Are Pure Luxury."

Graz 2003 achieved little through a master plan developed by town planners over many years for the (hopefully successful) turning point. Instead more was brought about by the pressure of deadlines punctiliously imposed by the city administration to follow the ideas of the programme director Lorenz so that the opening would not take place without spectacular events and institutions. This strait gave the director great influence but also prompted the city politicians to do some positive rethinking and eventually to identify with the projects. Alongside numerous artistic, literary and theatrical events, there is a full-scale explosion of projects in the architecture and public space sector which would never have been possible to realise without the pressure to succeed imposed by the capital. With gradually increasing pleasure in this development, Graz authorised "things the city needs but also things that are pure luxury" (Lorenz).
Building projects were realised deriving directly from the initiative of the programme plans as well as some by independent construction companies which were erected at the same time driven by the tail wind of the dynamic atmosphere of development all around. The range of buildings encompasses the Literaturhaus by Reigler Riewe, the

g2plus: Sparda-Bank

Riegler Riewe:
Literaturhaus

Riegler Riewe:
ÖBB-Kundenzentrum
im Hauptbahnhof Graz
Customer service centre
at Graz railway station

fasch & fuchs:
Kindermuseum
Children's museum

turhaus von Riegler Riewe, die Stadthalle von Klaus Kada, die „Listhalle“ von Markus Pernthaler, das Kindermuseum von fasch & fuchs, das Kunsthaus von Cook & Fournier (ein Joint Venture von Bund, Stadt und Land, ohne Beteiligung des Programmbudgets für Graz 2003), die von der Bundesbahn vorgezogene Bahnhofsrenovierung von Zechner & Zechner mit der Kunstintervention von Peter Kogler, die Murinsel von Vito Acconci, die Gestaltung der Stadteinfahrten durch mehrere junge Architektengruppen, die Umbauten am Flughafen von Markus Pernthaler, der ironische „Uhrturmschatten“ von Markus Wilfing, zahlreiche Großplastiken im öffentlichen Raum unter dem Titel „Concrete Art“, das Café „Graz 2003“ von Hans Gangoly und der „Marienlift“ bei einer barocken Mariensäule von Richard Kriesche. Und nun wurde auch das Volkskundemuseum der jungen Architektengruppe BEHF eröffnet.

Identifikationsfaktor Kultur

Diese stattliche Liste zeigt zumindest eine große Ambition der Organisatoren und auch eine große Begeisterung der Stadt, ihren Optimismus und ihren Glauben an eine neue Identität. Und diesmal sollte das Kulturprofil der Stadt nicht unabsichtlich entstehen, wie das in jenen seligen Avantgardejahren der Fall war, als Graz seine Kunstschaffenden eher als Plage empfand, die nur noch mehr ähnliche Plagegeister anzog. Sondern es wurde bewusst auf eine hochwertige kulturelle Infrastruktur gesetzt, mit der sich alle Grazer identifizieren sollten und die zumindest eine Chance auf zukünftige Plattform weiterer Entwicklungen bot. Programmintendant Lorenz sieht denn auch einen gelungenen Identitätswandel: „Graz hat in den beiden letzten Jahrzehnten, auch durch die politischen Zustände, eigentlich nicht mehr recht gewusst, was es mit sich machen soll. Die Stadt hat nicht an sich geglaubt, sie wusste nicht, ob sie sich musealisieren oder sich die Lust auf Zukunft, auf Avantgarde und Schräges erhalten soll, was in Graz ja immer interessant polarisiert war. Aber die Avantgarde war zuletzt nicht mehr die Avantgarde, da ihre Gegenpole nicht mehr funktioniert haben. Daher war das eine desperate Stimmung. Graz hat aber heute nicht nur einen Identitätswechsel vorgenommen – diesen Eindruck hat man ja, wenn man dort ist, es pulsiert und die Stadt wirkt happy mit sich –, sondern es hat erstmals seit Jahrzehnten überhaupt eine Identität gefunden, und diese Identität ist

Stadthalle by Klaus Kada, the Listhalle by Markus Pernthaler, the Kindermuseum (Children's Museum) by fasch & fuchs, the Kunsthaus by Cook & Fournier (a joint venture of the federal, city and provincial governments, without a contribution from the programme budget for Graz 2003), the Federal Railways renovation of the railway station by Zechner & Zechner with the artistic intervention by Peter Kogler, the Murinsel by Vito Acconci, the design of access roads to the city by several groups of young artists, conversions at the airport by Markus Pernthaler, the ironic Uhrturmschatten (clock tower shadow) by Markus Wilfing, numerous large sculptures in public spaces under the title Concrete Art, the Café Graz 2003 by Hans Gangoly and the Marienlift next to a baroque Mary column by Richard Kriesche. And recently BEHF's Folk Art Museum was opened.

Culture As A Factor Of Identification

This impressive list at least testifies to the great ambitions of the organisers and the considerable enthusiasm of the city, its optimism and belief in a new identity. This time the cultural image of the city should not come about by chance, as was the case during those blissful avant-garde years when Graz considered its artists a plague who only attracted more pests. Instead importance was placed on a cultural infrastructure of the highest quality with which the whole population of Graz should identify and which would, at the very least, provide the opportunity of a future platform for further development. Programme director Lorenz also sees a successful transformation of identity in this: "Partly due to political circumstances, Graz has not really known what to do with itself during the past couple of decades. The city did not believe in itself, it did not know whether it should turn itself into a museum or maintain an inclination for the future, for the avant-garde and unconventional, which was always interestingly polarised in Graz. However, the avant-garde was, in the end, no longer the avant-garde as its counterparts no longer functioned. That is why the mood was desperate. Graz has not only undertaken a change in identity, one has this impression when there, the city hums and seems happy with itself – but it has found an identity for the first time in decades, and this identity can be described with a relatively trivial word: pride. Graz is proud of itself, the citizens are proud of their city, and one no longer talks contin-

Kunstinterventionen am Flughafen und auf Zügen
Artistic intervention at the airport and on trains

Richard Kriesche:
Marienlift

Markus Wilfing:
Uhrturmschatten
Clock tower shadow

Hans Kuppelwieser:
Concrete Art

heute mit einem relativ trivialen Begriff zu beschreiben: Stolz. Graz ist stolz auf sich, die Bürger sind stolz auf ihre Stadt, und man redet nicht mehr ununterbrochen darüber, wie scheußlich es hier ist und wie rasch man hoffentlich von hier wegkommt, wie schön es jenseits des Semmerings oder an der Adria ist. Das ist eine maximale Leistung einer ‚Kulturhauptstadt', es war wichtig, mit der Gesellschaft zu arbeiten. Jeder muss das Gefühl haben, dass er Teil der Kulturhauptstadt ist."

Aber warum setzte die Stadt überhaupt auf Kultur, was erwartete man davon und was verstand man begrifflich darunter, als man sich um den „Hauptstadt"-Titel bewarb? Lorenz: „Kulturhauptstadtjahr ist ein Mehrwert, der nichts ersetzt oder abschafft, aber etwas anschafft, das zwar möglicherweise im gleichen Jahr aufgegessen wird, aber für die Zukunft ein Signal darstellt. Gesellschaftspolitik und Stadtentwicklung, das sind die zwei Karten, auf die wir gesetzt haben. Es heißt ja Kulturhauptstadt und nicht Kunsthauptstadt. Diese beiden Dinge – Gesellschaftspolitik und Stadtentwicklung – sind die einzigen, die Chance auf Nachhaltigkeit haben: Nämlich dass sich eine Gesellschaft über Kultur entwickelt und dass sich eine Stadt durch städtebauliche oder infrastrukturelle Maßnahmen weiterentwickelt und sich selbst eine Chance auf Zukunft gibt." Und inhaltlich: „Kultur ist die einzige Möglichkeit, eine Gesellschaft auch auf ein vereintes Europa vorzubereiten, das nur über die Kultur funktioniert und nicht über die Wirtschaft und nicht über die Politik. Aber was ist Kultur? Die Definition ist: So breit wie möglich, jeden bei sich selbst abholen, jeder hat Kultur, nur die wenigsten wissen es." Das Motto „für jeden etwas" scheint tatsächlich der Schlüssel für eine breite Identifikation der Bevölkerung mit dem Gebotenen zu sein, das dadurch zu einem Erarbeiteten wird.

Vito Acconci's Event-Arena

Und in dieser Unbekümmertheit entstanden fast beiläufig oder zufällig die städtebaulich interessanten Kunst-Verdichtungen. Eher intuitiv scheint im Intendanten die Erkenntnis gereift zu sein, dass die Mur, jener Fluss, der die Stadt in zwei ungleichwertige Hälften teilt, das Gelenk und der Brennpunkt einer integrierenden Neudefinition sein muss. Das Angebot des New Yorker Universalkünstlers Vito Acconci, eine künstliche Insel im Fluss zu errichten, lag bereits längere Zeit vor, als Lorenz plötzlich sein Potenzial erkannte. Mit dieser räumlichen Verbindung zwischen den traditionellen Kulturstätten der Innenstadt und der eher gewerblich-industriell orientierten „anderen" Flussseite konnten – neben der gelungenen Stadtstimulation – gleich mehrere Effekte erzielt werden. Das neue Kunsthaus, ein regionales Museum moderner Kunst, das auf eine lange und traumatische Planungszeit in mehreren Wettbewerben an immer wieder verworfenen Standorten zurückblickt, war bereits unter dem Erfolgs- und Termindruck des Hauptstadtjahrs auf dieser „anderen" Seite angesiedelt worden. Gleich daneben befindet sich schon seit langem ein weiteres Kunstzentrum: Die Veranstaltungsräume der Minoriten, die hier seit Jahrzehnten einen gut eingeführten Kunstplatz betreiben. Diese beiden Häuser ließen sich nun durch die Insel nicht nur gut mit der am anderen Flussufer liegenden Altstadt verbinden. Das muschelförmig-bizarre Stahlobjekt von Acconci, das halb eingeschlossener

uously about how horrible it is here and how quickly one can get away, how beautiful it is beyond Semmering or on the Adriatic coast. This is a great achievement for the cultural capital, it was important to work with society. Everyone has to have the feeling that they are part of the cultural capital." But why did the city rely on culture at all? What did one expect from it and what did one understood by it when competing for the title of capital? Lorenz: "Cultural Capital year is of increment value, neither replacing nor abolishing anything but bringing something in which, although it may well be used up in the same year, represents a signal for the future. Social politics and urban development, we have put everything on these two cards. The phrase is cultural capital and not art capital. Both of these things, social politics and urban development, are the only ones with any chance of lasting: that is to say, a society develops through culture and a city further develops through the measures of urban development and infrastructure, giving itself a chance for the future." And concerning content: "Culture is the only possibility to also prepare a society for a unified Europe – which only functions through culture, not through economics or politics. But what is culture? The definition is: as broad as possible, everybody should claim it for themselves, everyone has culture, only the few know it though." The motto 'something for everybody' seems to be the key to a broad identification in the population with what is on offer, becoming as a result hard-won.

Vito Acconci's Event-Arena

Interesting artistic augmentations to urban development are produced almost incidentally or accidentally in this carefree atmosphere. It appears that the director intuitively recognised that the Mur, the river which divides the city into two unequal halves, has to be the link and the focal point of an integrating new definition. The offer made by the New York universal artist Vito Acconci to create an artificial island had existed for sometime before Lorenz suddenly recognised its potential. With this connection between the traditional cultural sites of the inner city and the business and industry oriented other side of the river, several effects could be achieved simultaneously besides successfully stimulating the town. The new Kunsthaus, a regional museum of modern art which had experienced a long and traumatic period of planning in several competitions at repeatedly rejected sites, had already been located on the other side under the pressure to meet deadlines and succeed. Another art centre has existed right next to it for some time: the function rooms of the Minor Friars, who have run a well-received art venue here for decades. Both of these buildings were now readily connected to the old town on the other bank via the island. The bizarre, shell-shaped steel object by Acconci, one half being an enclosed space and the other a hollowed-out event arena, defines its own space. It occupies the river, which had previously functioned as a no-mans-land and caesura, and generates new routes in its environs, a revaluation of its surroundings and new axes of perspective. "Nobody can imagine today that this island will be dismantled", says Lorenz, "it is a part of the heart of Graz 2003, beating strongly. We have had hundreds of thousands of visitors on this island, it is always full."

Hans Gangoly: Café Graz 2003

Raum, halb ausgehöhlte Event-Arena ist, definiert auch seinen eigenen Raum, besetzt den bislang als Niemandsland und als Zäsur fungierenden Fluss, und es generiert in seinem Umfeld neue Wege, eine Aufwertung des Umfeldes und neue Sichtachsen. „Es ist heute unvorstellbar für alle, dass diese Insel wieder demontiert wird", sagt Lorenz, „es ist ein Herzstück von Graz 2003, das kräftig schlägt, wir haben hunderttausende Besucher auf dieser Insel gehabt, es ist immer voll." Soviel Leistung konnte man von einem derartigen Objekt, das von vielen zunächst nur als schwimmendes Kunstobjekt gelesen wurde, kaum erwarten. „Ob es an der Mur Promenaden gibt oder nicht," meint Lorenz, „ist im Gesamtkontext des neuen Bezirks am Fluss, der nun entsteht – mit Kunsthaus, Murinsel, Murpromenaden und anderem – sehr wichtig und es ist wesentlich, den Fluss ergehen, begehen und begleiten zu können. Es ist einfach passiert, dass hier eine Stadt nach innen wächst und sich sozusagen einen eigenen Bezirk herausschneidet, nämlich den Bezirk an der und um die Mur. Dieses Bewusstsein, den Fluss wieder zu gewinnen und eine Gleichwertigkeit zwischen den Stadtteilen herzustellen, hat viel mit Stadt- und Gesellschaftspolitik und mit Befindlichkeit in der Stadt zu tun. Das hat die Stadt, wie ich meine, sehr verändert und das ist etwas, was selbst die schlechteste Nachfolgepolitik nicht mehr verhauen kann – das ist, und das bleibt vom Kulturhauptstadtjahr und das freut mich auch."
Der nun entstandene Bereich bietet in der Tat die Chance auf Nachhaltigkeit, da er von den Bewohnern bereits jetzt auf selbstverständliche Art genutzt wird. Zwischen Kunsthaus, Minoriten, Café „Graz 2003", dem Platz davor, dem spitzwinklig-expressiven Mursteg von Günther Domenig, der organoid-expressiven Murinsel von Acconci und ihren zuführenden Stegen sowie den anschließenden Promenaden und dem nahen „Domenig"-Spitz (einer städtebaulich spannenden Eckbebauung) ist tatsächlich Neuland entstanden. Lorenz: „In diesem Bezirk innerhalb der Stadt ist plötzlich ein neues Stadt- und Naturleben entstanden, von dem wir eigentlich nicht einmal zu träumen gewagt hätten. Faktisch hat sich hier etwas verclustert und es ist ein neuer Stadtbezirk daraus geworden."

So much could hardly have been expected of an object which was originally seen by many only as a floating artwork. "Whether there are promenades along the Mur or not", says Lorenz, "it is very important and essential for the whole concept of the new district now being created on the river bank with the Kunsthaus, Mur Island, Mur promenades etc., to be able to stroll along the river, to access and accompany it. It simply happened that a city grew into itself here and cut out a distinct district from itself, namely the district along and around the Mur. This consciousness of having won back the river and creating a parity between both parts of the town has a lot to do with city and social politics and with the state of being in the city. This has changed the city a lot in my opinion, and that is something which not even the worst future political constellation can wreck, something that is and that will remain from the year of being cultural capital, and that I'm glad about that, too."
The area now created really does offer the chance for something more lasting because it is already being used by the residents as a matter of course. New land has really been created between the Kunsthaus, the Minor Friars, Café Graz 2003, the square in front of it, the sharp-angled, expressive Mursteg by Günther Domenig, the organoid-expressive Mur Island by Acconci and the bridges leading to it, as well as the promenades and the nearby Domenig-Spitz (a building representing an exciting piece of urban development). Lorenz: "In this district inside the city, new town and natural life have suddenly been created, which we would barely have dared dream of. Something has clustered together here and a new city district has been produced."

Vito Acconci:
Murinsel
Island in the river Mur

Architektonische Aufrüstung

Der Bezirk auf und um die Mur ist indes nur *ein* Schauplatz der zahlreichen Aufrüstungen von Graz. Ein weiterer wichtiger Schwerpunkt ist etwa die Aufwertung der Verkehrsinfrastruktur durch bauliche und künstlerische Interventionen. Markus Pernthaler hat einen neuen, anspruchsvollen Flughafentower gebaut, in der Abflughalle wurde eine Wandgestaltung von Flora Neuwirth realisiert, am Bahnhof, dessen Umbau innerhalb des „Bahnhofsoffensive"-Programms der Bundesbahn durchgeführt wurde, entstand ein Riesenmural von Peter Kogler, an den Stadteinfahrten gaben die jungen Archietktengruppen PENTAPLAN, ORTLOS architects, [x architekten] und LOVE Kostproben ihres Könnens und Humors.
Die wichtigen neuen Institutionen wie „Listhalle", Kunsthaus, Literaturhaus und Kindermuseum versprechen zumindest starke Stimulationen. Die Listhalle von Markus Pernthaler – eine auf akustisch international höchstwertigen Standard aufgewertete und umgebaute ehemalige Industriehalle – bietet die Bühne für Hi-End-Musikveranstaltungen, das bubbleförmige Kunsthaus des alternden Pop-Art-Architekten Peter Cook die Plattform spektakulärer Inszenierungen aktueller Kunst und das neue Literaturhaus kann ja bereits an eine große lokale Tradition anknüpfen. Das zweite Kindermuseum Österreichs (das erste entstand im Wiener MuseumsQuartier) sollte bald Vorgabe für das Standardrepertoire mittelgroßer europäischer Städte sein.
Eine wesentliche Dimension aller Kulturbauten für ihre Funktionsfähigkeit im Stadtganzen jedoch hat Lorenz deutlicher erkannt als mancher Politiker und mancher Architekt, nämlich die sinnvolle Nutzung, die einerseits planerisch ermöglicht, andererseits aber vor allem auch durch Identifikation dauerhaft gemacht werden muss: „Wenn die Gesellschaft auf das Bauen nicht reagiert und die Häuser nicht kommunikationstauglich sind, dann haben wir Kunstruinen oder Architekturruinen, was ich natürlich keinem der in Graz entstehenden Häuser wünsche. Aber wir haben das schon erlebt. Wenn Sie sich beispielsweise andere Städte ansehen, seitdem vor dreißig Jahren die Kulturbauten-BSE ausgebrochen ist – ich will gar nicht sagen, was da alles gebaut worden ist, bis hin zu Mönchengladbach, das heute im Grunde eine Institution in höchster Krise ist und einmal DAS Vorzeigeprojekt für neuen Museums- und Kunsthausbau war. Häuser zu bauen ist das eine, sie zu betreiben und sie für die Gesellschaft kommunikationsfähig zu machen ist das andere. Der Schlüssel zu allem ist immer der Mensch, und nicht die gebaute Substanz."

An Architectonic Upgrade

The district on and along the Mur is only one of the showplaces of the numerous examples of 'upgrading' in Graz. An important emphasis has, for example, been laid on the revaluation of the transport infrastructure through structural and artistic interventions. Markus Pernthaler has built a new and demanding airport control tower, a wall design by Flora Neuwirth was realised in the departure hall, at the railway station, renovated as part of the Federal Railways Bahnhofsoffensive (station offensive) programme, an enormous mural was created by Peter Kogler, and the groups of young architects PENTAPLAN, ORTLOS architects, [x architekten] and LOVE presented samples of skill and humour.
The important new institutions, such as the Listhalle, Kunsthaus, Literaturhaus and Kindermuseum, promise to be extremely stimulating at the very least. The Listhalle by Markus Pernthaler, a former factory hall converted and brought up to the highest international standards for acoustics, provides a stage for hi-end music performances, the bubble-shaped Kunsthaus by the aging pop art architect Peter Cook provides a platform for spectacular shows of contemporary art, and the new Literaturhaus can pick up on a great local tradition already in existance. The second Austrian Kindermuseum (the first was created in the Vienna MuseumsQuartier) will apparently soon become the blueprint for the standard repertoire of medium-sized European cities.
An important dimension governing the ability of all cultural buildings to function in the city as a whole, however, was more clearly recognised by Lorenz than by some politicians and architects, namely sensible use, which must be made possible on the one hand by planning and, on the other, made sustainable through identification above all else: "When society does not react to building and the buildings are not able to communicate, we have art ruins or architectural ruins, something I would not of course wish for any of the buildings being erected in Graz. But we have already experienced it. If you look at other cities, for example, ever since the BSE of cultural buildings broke out – I don't want to talk about what was built right up to Mönchengladbach, today an institution experiencing a major crisis but one which was THE show-piece project for new museum and exhibition hall architecture. To erect buildings is one thing, to run them and to make them able to communicate with society is another. The key to everything is always the people and not the building itself."

Stadteinfahrten von | City access by **PENTAPLAN und** | and **ORTLOS Architekten**

Baustelle Kunsthaus Graz
Construction site Kunsthaus Graz
Photo Andreas Haller

Raimund Abraham
geboren 1933 in Lienz/Osttirol. Architekturstudium an der Technischen Universität in Graz. Lebt und arbeitet seit 1964 in den USA. 1971-2002 Professor für Architektur an der Cooper Union in New York. Als Exponent der Wiener Avantgarde der frühen 1960er Jahre wurden seine Arbeiten zusammen mit denen von Hans Hollein und Walter Pichler bereits 1967 im Museum of Modern Art, New York ausgestellt. Abraham baute u.a. in Berlin (IBA-Wohn- und Geschäftshaus, 1985), in Wien (Siedlung Traviatagasse, 1991), in Graz (Wohnhaus, 1993), in Lienz (Hypobank, 1998), in New York (Anthology Film Archives, 1989 und das Austrian Cultural Forum, 2002) und sein eigenes Haus in Mexiko sowie für den Mäzen und Immobilienmakler Karl Heinz Müller auf einem ehemaligen Militärgelände bei Neuss in Nordrhein-Westfalen ein Musikerhaus im Rahmen des Projekts Museumsinsel Hombrioch.

born 1933 in Lienz/East Tyrol. Studied architecture at Graz University of Technology. Has been living and working in the USA since 1964. 1971-2002 architecture professor at the Cooper Union in New York. As a proponent of the Viennese avant-garde in the 1960s, his work was exhibited along with that of Hans Hollein and Walter Pichler in the Museum of Modern Art, New York. Abraham has, among other projects, completed buildings in Berlin (IBA apartment and commercial building, 1985), in Vienna (Traviatagasse housing estate, 1991), in Graz (apartment building, 1993), in Lienz (Hypobank, 1998), in New York (Anthology Film Archives, 1989 and the Austrian Cultural Forum, 2002), his own house in Mexico and is currently designing a musicians' house as part of the Hombrioch Museumsinsel project.

Ernst Beneder
geboren 1958 in Waidhofen/Ybbs, Architekturstudium an der TU Wien, Postgraduate Studien am Tokyo Institute of Technology, fellow der Japan Society for the Promotion of Science.
Seit 1987 Architekt in Wien, seit 1996 Arbeitsgemeinschaft mit Anja Fischer. Gastprofessuren an der TU Wien und an der University of Illinois. 2001 erhielt er den Otto Wagner-Städtebaupreis für das Stadtprojekt Waidhofen/Ybbs.
Realisierungen (Auswahl): Heimatmuseum Waidhofen (1999), Osterkapelle Stift Herzogenburg (2000), Sparkasse Niederösterreich St. Pölten (2000), Wohnhausanlage in Wien (2001), Stadtprojekt Waidhofen/Ybbs und das Otto Wagner Spital/Neurologisches Zentrum (2002).

born in 1958 in Waidhofen an der Ybbs/Austria, studied architecture at Vienna University of Technology, post-graduate studies at the Tokyo Institute of Technology, fellow of the Japan Society for the Promotion of Science. Since 1987, freelance architect in Vienna, since 1996 collabortion with Anja Fischer. Visiting Professor at Vienna University of Technology and at the University of Illinois. 2001 he was awarded the Otto Wagner Urban Development Prize for the Waidhofen an der Ybbs urban project.
Realised projects (selected): Waidhofen regional folk museums (1999), Stift Herzogenburg Easter Chapel(2000), Lower Austria Sparkasse bank St. Pölten (2000), housing complex in Vienna (2001), the Waidhofen an der Ybbs urban project and the Otto Wagner Spital/neurological centre (2002).

Herzog & de Meuron

Jacques Herzog
geboren 1950 in Basel, Studium an der ETH Zürich (Diplom 1975). 1983 Gastprofessur an der Cornell University, New York; 1989 und seit 1994 Gastprofessur an der Harvard University, Cambridge; seit 1999 Professur ETH-Zürich, ETH-Studio in Basel.

born in 1950 in Basle/Switzerland, studied at Zurich ETH (graduated 1975).
1983 visiting professor at Cornell University, New York; 1989 and since 1994 visiting professor at Harvard University, Cambridge; since 1999 professor at the ETH Zurich, ETH-Studio in Basle.

Pierre de Meuron
geboren 1950 in Basel, Studium an der ETH Zürich (Diplom 1975). 1989 und seit 1994 Gastprofessur an der Harvard University, Cambridge; seit 1999 Professur ETH-Zürich, ETH-Studio in Basel.

born in 1950 in Basle/Switzerland, studied at Zurich ETH (graduated 1975). 1989 and since 1994 visiting professor at Harvard University, Cambridge; since 1999 professor at the ETH Zurich, ETH-Studio in Basle.

Harry Gugger
geboren 1956 in Grezenbach/Schweiz. Studium an der ETH Zürich und der Columbia University, New York (Diplom 1989). Seit 1990 Mitarbeiter bei Herzog & de Meuron, seit 1991 Partner. 2001 Gastprofessur EPF-Lausanne, Schweiz.

born in 1956 in Grezenbach/Switzerland, studied at Zurich ETH and Columbia University, New York (graduated 1989). Worked at Herzog & de Meuron since 1990, partner since 1991. 2001 visiting professor at the EPF Lausanne, Switzerland.

Christine Binswanger
geboren 1964 in Kreuzlingen/Schweiz. Studium an der ETH Zürich (Diplom 1990), seit 1991 Mitarbeit bei Herzog & de Meuron, seit 1994 Partner. 2001 Gastprofessur EPF Lausanne, Schweiz.

born in 1964 in Kreuzlingen/Switzerland, studied at Zurich ETH (graduated in 1990), worked at Herzog & de Meuron since 1991, partner since 1994. 2001 visiting professor at the EPF Lausanne, Switzerland.

Das Büro wurde 1978 von Jacques Herzog und Pierre de Meuron in Basel gegründet und hat sich in den letzten beiden Jahrzehnten zu einem der weltweit führenden Architekturbüros entwickelt.
Herzog & de Meuron können auf ein umfangreiches Werk verweisen, darunter das Wohn- und Bürohaus Schwitter in Basel, das Lagerhaus Ricola in Laufen, das Stellwerk in Basel, die Bibliothek in Eberswalde, das Kunstmuseum in Duisburg, die Tate Modern oder das Laban Centre for Dance in London. Das Büro wurde vielfach ausgezeichnet, u.a. 2001 mit dem Pritzker Preis und hat an zahlreichen internationale Ausstellungen teilgenommen.

The studio was founded in Basle in 1978 by Jacques Herzog and Pierre de Meuron, and in the last two decades it has developed to become one of the leading architecture studios in the world.
Herzog & de Meuron can look back over a comprehensive body of work which includes the Schwitter apartment and an office building in Basle, the Ricola warehouse in Laufen, the signalbox in Basle, the library in Eberswalde, the art museum in Duisburg, the Tate Modern or the Laban Centre for Dance in London. The practise has won many awards, including the Pritzker Prize in 2001, and participated in numerous international exhibitions.

Lab architecture studio
www.labarchitecture.com

Donald Bates
geboren 1958 in Texas, Architekturstudium an der University of Houston (Diplom 1978). Er arbeitete in verschiedenen Architekturbüros, u.a. bei Daniel Libeskind.

born in 1953 in Texas.
He studied architecture at the University of Houston (graduated in 1978) and worked in several studios, a.o. at Daniel Libeskind.

Peter Davidson
geboren in Newcastle/Australien, Architekturstudium am New South Wales Institute of Technology in Sydney (Diplom 1981).

born in Newcastle/Australia.
He studied architecture at the New South Wales Institute of Technology in Sydney (graduated in 1981).

1994 Gründung von Lab architecture studio, 1997 gewannen Donald Bates and Peter Davidson den internationalen Wettbewerb zum Federation Square in Melbourne.

In 1994 Donald Bates and Peter Davidson formed Lab architecture studio, in 1997 they won the international design competition for the Federation Square in Melbourne.

Dominique Perrault
geboren 1953, Architekturstudium an der Ecole Supérieure des Beaux-Arts in Paris (Diplom 1978) und an der Ecole Supérieure des Ponts et Chaussées Stadtplanung (Diplom 1979). 1980 Postgraduate Studium der Geschichte an der Ecole des Hautes Etudes en Sciences sociales. 1981 Eröffnung des eigenen Büros in Paris. Perrault hat zahlreiche Wettbewerbe gewonnen, darunter 1989 den Wettbewerb für die Französische Nationalbibliothek in Paris – sein berühmtester Bau – und 1991 für Sporteinrichtung für die Bewerbung Berlins für Olympischen Spiele. Die Radsporthalle wurde 1997 fertiggestellt, die Schwimmsporthalle 1999, das Rathaus mit Büros und Einkaufspassage in Innsbruck 2002. Zur Zeit arbeitet das Büro Perrault an folgenden Projekten: Erweiterung des Europäischen Gerichtshofs in Luxemburg, Sportkomplex in Madrid, Masterplan und Hochhaus Donau City Wien, Fußballstadion in Montigalà in Badalona/Spanien, zwei Hotelhochhäuser in Barcelona, Fabrik für Metallgewebe in Maryland/USA, Bau von mehreren M-Preis-Märkten in Österreich, ein städtebauliches Projekt für Teneriffa und eine Platzgestaltung in Mailand.

was born in 1953 and studied architecture at the Ecole Supérieure des Beaux-Arts in Paris (graduated 1978) and town planning at the Ecole Supérieure des Ponts et Chaussées (graduated 1979). In 1980 he completed a course of post-graduate studies in history at the Ecole des Hautes Etudes en Sciences Sociales. In 1981 he opened his own office in Paris. Perrault has won numerous competitions, among them the French National Library in Paris – his best known building – in 1989 and, in 1991, the cycling and swimming stadium in Berlin. The cycling stadium had been completed in 1997, the swimming stadium in 1999. The town hall with offices and shopping passage in Innsbruck was completed in 2002. Perrault's studio is currently working on the following projects: extension to the European Court in Luxembourg, sports complexe in Madrid, Donau City masterplan and high-rise in Vienna, Montigalà football stadium in Badalona/Spain, two hotels in Barcelona, factory for metalwork in Maryland/USA, construction of several M-Preis supermarkets in Austria, an urban development project in Tenerife and a design for a public square in Milan.

Billie Tsien
Studium in Yale und Architekturstudium an der UCLA. Lehrtätigkeit am Southern California Institute of Architecture in L.A., an der Parson School of Design in New York, in Yale, Harvard, University of Texas in Austin und an der University of Pennsylvania. Kürzlich teilte sich Billie Tsien mit Tod Williams den Louis Kahn Lehrstuhl an der Yale University. Seit 1977 Zusammenarbeit mit Tod Williams mit Büro in New York.

Studied at Yale and at the UCLA. She has taught at the Southern California Institute of Architecture in L.A., at the Parson School of Design in New York, at Yale, Harvard, University of Texas in Austin and at the University of Pennsylvania.
She recently shared the Louis Kahn chair at the Yale University with Tod Williams.
She has been working in collaboration with Tod Williams since 1977 with a studio in New York.

Tod Williams
Architekturstudium in Princeton und an der Cambridge University. Lehrtätigkeit an der Cooper Union in New York, an der University of Texas in Austin und an der Yale University. Seit 1977 Zusammenarbeit mit Billie Tsien mit Büro in New York.

Studied architecture at Princeton and at the Cambridge University. He has taught at the Cooper Union in New york, at the University of Texas in Austin an at Yale University. He has been working in collaboration with Billie Tsien since 1977 with a studio in New York.

Realisierungen (Auswahl):
The New College, University of Virginia, Phoenix Art Museum und Theater in Phoenix/Arizona, The Neurosciences Institutes in La Jolla/Californien und The Cranbrook Athletic Complex in Bloomfield Hills/Michigan, für die Cooper Union in New York ein Computer Center und das American Folk Art Museum in New York.

Realised projects (selected):
The New College, University of Virginia, Phoenix Art Museum and Theater in Phoenix/Arizona, The Neurosciences Institutes in La Jolla/Californien and The Cranbrook Athletic Complex in Bloomfield Hills/Michigan, a computer centre for the Cooper Union in New York and the das American Folk Art Museum in New York.

Harald Bodenschatz
geboren 1946; Stadtsoziologe und Stadtplaner, seit 1995 Professor für Planungs- und Architektursoziologie an der TU Berlin. Planerische Praxis seit 1980 in der Stadterneuerung. Längere Lehr- und Forschungsaufenthalte in Italien, Brasilien, USA, Peru, Argentinien und England.
Seit 1993 Mitglied der Redaktion der Zeitschrift „Die alte Stadt". Preisträger des Verbands der deutschen Kritiker e.V. für die Sparte Architektur (zusammen mit Stephan Braunfels) 1994. Mitglied des AIV, BDA, SRL, der Deutschen Akademie für Städtebau und Landesplanung sowie des Deutschen Werkbundes Berlin. Mitglied des „Schinkelzentrums" der TU Berlin. Zahlreiche Veröffentlichungen zur Stadterneuerung, zum postmodernen Stadtumbau, zum suburbanen Städtebau sowie zur Stadtplanungs- und Stadtbaugeschichte.

born in 1946; urban sociologist and planner, planning and architecture sociology professor at Berlin University of Technology since 1980. Planning practise since 1980 in the urban renewal sector. Long teaching and research trips to Italy, Brazil, USA, Peru, Argentina and England. On the editorial team of the journal Die Alte Stadt since 1993. Verband der Deutsche Kritiker (Society of German Critics) 1994 prize for architecture (with Stephan Braunfels). Member of the AIV, BDA, SRL, the German Academy of urban development and regional planning as well as the Deutsche Werkbund in Berlin. Member of the Schinkelzentrum at Berlin University of Technology. Numerous publications on urban renewal, postmodern urban development, suburban development, and on the history of urban planning and development.

Matthias Boeckl
geboren 1962 in Wien, Studium der Kunstgeschichte, Promotion 1988 an der Universität Wien, Habilitation 1999 an der Universität Innsbruck. Professor für Geschichte und Theorie der Architektur an der Universität für angewandte Kunst Wien.
Seit 1999 Chefredakteur von „architektur. aktuell" (SpringerWienNewYork).
Autor zahlreicher Aufsätze und Bücher über Themen der modernen und zeitgenössischen Kunst und Architektur.

born in 1962 in Vienna; studied art history, Ph.D. at Vienna University in 1988. Architecture history and theory professor at Vienna University of Applied Arts. Editor in chief of architektur.aktuell (SpringerWienNewYork) since 1999. Author of numerous texts and books on modern and contemporary art and architecture.

Cordelia Polinna
geboren 1975; Studium der Stadt- und Regionalplanung an der Technischen Universität Berlin sowie Town and Country Planning am Edinburgh College of Art, Wissenschaftliche Mitarbeiterin bei dem von der DFG geförderten Forschungsprojekt „Postmoderner Städtebau seit den 80er Jahren: Remake der ‚Europäischen Stadt' am Beispiel von Berlin und London" unter der Leitung von Prof. Harald Bodenschatz, Fachgebiet Planungs- und Architektursoziologie, TU Berlin (2002/03) währenddessen Forschungsaufenthalte in London, zur Zeit Promotion.

born in 1975; studied urban and regional planning at Berlin University of Technology as well as town and country planning at Edinburgh College of Art, research assistant on the DFG-supported research project Postmoderner Städtebau seit den 80er Jahren: Remake der Europäischen Stadt am Beispiel von Berlin und London (Postmodern Urban Development Since the 1980s: Remake of the European City Based on the Examples of Berlin and London) under the leadership of Prof. Harald Bodenschatz, specialist for planning and architecture sociology at Berlin University of Technology (2002/03). Research trip to London during this period.
She is currently working on her thesis.

Viviane Stappmanns
1976 in Wiesbaden geboren, kam 1996 zum ersten mal nach Melbourne, wo sie in den folgenden Jahren Journalistik studierte. Heute arbeitet sie als Editorial Coordinator der australischen Design-Zeitschrift (inside) interior review, und als freie Journalistin für internationale Architektur- und Design-Publikationen. Weiterhin unterrichtet sie am Royal Melbourne Institute of Technology (RMIT) im Fachbereich Interior Design, und absolviert mit einem Forschungsprojekt zum Thema Design-Journalismus einen Master of Arts in Interior Design.

born in 1976 in Wiesbaden; first went to Melbourne in 1996, where she studied journalism for the ensuing years. She is currently working as editorial coordinator of the Australian design magazine (inside) interior review and as a freelance journalist for architecture and design publications around the world. She also teaches in the field of interior design at the Royal Melbourne Institute of Technology (RMIT) and is currently completing an MA with a research project on design journalism.

Wolfgang Jean Stock
geboren 1948. Studium von Neuerer Geschichte, Politischer Wissenschaft und Spziologie. Berufliche Hauptstationen: 1978 bis 1985 Direktor des Kunstvereins München, 1986 bis 1993 Architekturkritiker der Süddeutschen Zeitung, 1994 bis 1998 stv. Chefredakteur der Zeitschrift „Baumeister". Seither freier Journalist und Buchautor in München. Letzte Buchveröffentlichung: Europäischer Kirchenbau 1950-2000, München/Berlin/London/New York 2002. Außerordentliches Mitglied im Bund Deutscher Architekten (BDA), Landesverband Bayern.

born in 1948; studied recent history, political science and sociology. Career highlights: 1978-85, director of the Munich Kunstverein, 1986-93 architecture critic for the Süddeutsche Zeitung, 1994-98 acting editor in chief of the journal Baumeister. Freelance journalist and author based in Munich. Most recently published book: Europäischer Kirchenbau 1950-2000 (European Church Building), Munich/Berlin/London/New York 2002.
Member of the Bavarian branch of the Bund Deutscher Architekten (BDA).

Klaus-Dieter Weiß
www.klausdieterweiss.de

geboren 1951 in Weimar, freier Autor und Publizist, lebt in Minden/Westfalen; Studium in München und Aachen, Forschungs- und Lehrtätigkeit am Institut für Entwerfen und Architektur der TU Hannover; Korrespondent von architektur.aktuell und „Werk, Bauen + Wohnen", zahlreiche Buchveröffentlichungen und Werkmonographien, zuletzt: „Zoom. Bothe Richter Teherani", Basel/Boston/Berlin 2003.

born in 1951 in Weimar; freelance author and journalist, lives in Minden/Westfalen. Studied in Munich and Aachen, research and teaching post at the Institute for Design and Architecture at Hanover University of Technology; correspondent for architektur.aktuell and Werk, Bauen + Wohnen; he has published numerous books and monographs on projects, most recently: Zoom. Bothe Richter Teherani, Basel/Boston/Berlin 2003.

SpringerArchitektur

Udo Kultermann

Die Architektur im 20. Jahrhundert

Sechste, überarbeitete und erweiterte Auflage.
2003. X, 355 Seiten. 319 Abbildungen.
Format: 15 x 20,5 cm
Broschiert **EUR 29,80,** sFr 48,–
ISBN 3-211-83887-2

Die von Udo Kultermann verfasste „Architektur im 20. Jahrhundert“ ist einmalig in ihrer Kürze und Präzision. Als Hand- und Lehrbuch für Studenten aber auch für Architekten und interessierte Laien, bietet es einen Überblick über die Fülle der verschiedenen Bauten und Baustile des 20. Jahrhunderts. In die Betrachtungsweise des Autors fließen, neben der Baukunst im engeren Sinne, auch eine analytische Darstellung des Städte- und Ingenieurbaus mit ein. Der bewährte Klassiker liegt hiermit in seiner 6. Auflage vor und wurde vom Autor aktualisiert und um ein Kapitel erweitert.

Aus dem Inhalt:

Puristische Tendenzen
Architektur, Wissenschaft und Industrie und die Einstellung zur Maschine

Empiristische Tendenzen
Organische Architektur und Technik im Dienst des Menschen und der Natur

Synkretistische Tendenzen
Städtebauliche Architektur und Bauen in Zusammenhängen

Autonome Architektur seit 1970
Regionale Identität und die Wiedergewinnung von Tradition

„... ein hilfreicher Wegbegleiter durch die Bauweisen unseres Jahrhunderts ..“

Die Welt

„Selten ist über so heterogene Dinge so intelligent geschrieben worden ...“

Heidelberger Tageblatt

SpringerWienNewYork

A-1201 Wien, Sachsenplatz 4–6, P.O. Box 89, Fax +43.1.330 24 26, e-mail: books@springer.at, Internet: www.springer.at
Birkhäuser c/o SAG, D-69126 Heidelberg, Haberstraße 7, Fax: +49.6221.345-4229, e-mail: orders@springer.de
Chronicle Books, USA, San Francisco, CA 94105, 85 Second Street, Fax +1.800.858-7787, e-mail: sales@papress.com

SpringerArchitektur

Matthias Boeckl (Hrsg.)

MuseumsQuartier Wien

Die Architektur / The Architecture

2001. 144 Seiten. Zahlreiche farbige Abbildungen.
Format: 21 x 29,7 cm
Text: deutsch/englisch
Gebunden **EUR 28,90**, sFr 46,50
ISBN 3-211-83641-1
Edition Architektur aktuell, Bd. 3

Mitten im Zentrum Europas entstand einer der zehn größten Kulturkomplexe der Welt.
Die spektakuläre Verbindung zwischen historischer (Fischer von Erlach) und moderner Architektur (Ortner * Ortner) ist die Form gewordene Vision eines Kulturviertels der Superlative. Das Buch zeigt nicht nur die neue und alte (Museums-) Architektur, sondern auch die Auswirkungen des gesamten Vorhabens auf das Stadt- und Kulturleben Wiens.

Mit Beiträgen u.a. von Friedrich Achleitner, Dieter Bogner, Bart Lootsma, Erich Raith und Matthias Boeckl.

„... Das Buch verschafft einen wunderbaren Einstieg in die jüngste städtebauliche Facette Wiens und ist allen an Wien interessierten Lesern empfohlen." Konstruktiv

„Man muss das zweisprachige Buch ‚Museumsquartier Wien – Die Architektur/The Architecture' lesen ... Das Buch enthält so viel kaum bekanntes historisches Material, dass es auch als Viennensium ein Muss ist ..." Die Furche

SpringerWienNewYork
A-1201 Wien, Sachsenplatz 4–6, P.O. Box 89, Fax +43.1.330 24 26, e-mail: books@springer.at, Internet: www.springer.at
Birkhäuser c/o SAG, D-69126 Heidelberg, Haberstraße 7, Fax: +49.6221.345-4229, e-mail: orders@springer.de
Chronicle Books, USA, San Francisco, CA 94105, 85 Second Street, Fax +1.800.858-7787, e-mail: sales@papress.com

SpringerArchitektur

Ákos Moravánszky (Hrsg.)

Architekturtheorie im 20. Jahrhundert

Eine kritische Anthologie

Unter Mitarbeit von Katalin M. Gyöngy.
2003. VIII, 591 Seiten. 108 Abbildungen.
Format: 16,5 x 24,2 cm
Gebunden **EUR 59,80,** sFr 96,–
ISBN 3-211-83743-4

Architekten, Publizisten, Kunsthistoriker, sie alle kennen es – das Vergnügen und die Qual, die Entwurfsideen der gesichteten oder ersonnenen Projekte vor einem kritischen Publikum zu erklären. Denn es sind die Wörter, die die Architektur am Leben erhalten und die Entscheide des Entwerfenden im Diskurs vermitteln.
Mit dieser einzigartigen Textanthologie der „Architekturtheorie im 20. Jahrhundert" liegt nun ein hilfreiches Vademekum vor. Einhundert Quelltexte von Architekten, Künstlern, Kunsthistorikern und Philosophen stehen dem Leser in kommentierter Form für die eigenen Recherchen – nach fünf Themen geordnet – in handlicher Form zur Verfügung. Alle fünf Themen werden mit Essays des Herausgebers eingeleitet, die 100 Texte sind jeweils zum tieferen Verständnis kommentiert.

Aus dem Inhalt:

Vom Stilus zum Branding
Die Wahrnehmung des Raumes
Konstruktionen der Natur
Monumentalität
Der Ort der Architektur

SpringerWienNewYork

A-1201 Wien, Sachsenplatz 4–6, P.O. Box 89, Fax +43.1.330 24 26, e-mail: books@springer.at, Internet: www.springer.at
Birkhäuser c/o SAG, D-69126 Heidelberg, Haberstraße 7, Fax: +49.6221.345-4229, e-mail: orders@springer.de
Chronicle Books, USA, San Francisco, CA 94105, 85 Second Street, Fax +1.800.858-7787, e-mail: sales@papress.com

SpringerArchitektur

Erich Raith

Stadtmorphologie

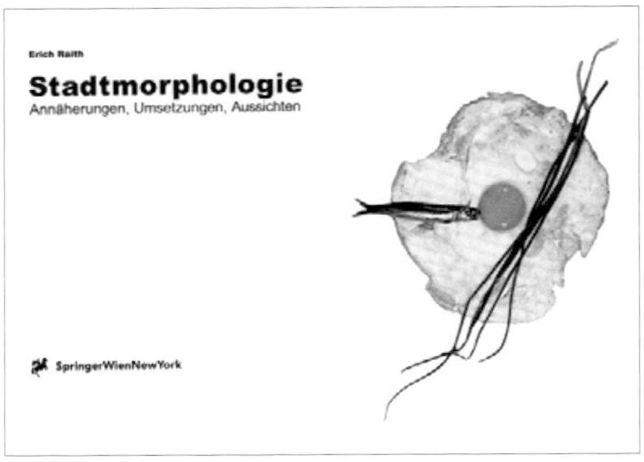

2000. II, 215 Seiten. Zahlr., z. T. farb. Abb.
Format: 29,7 x 21 cm
Broschiert **EUR 35,80,** sFr 57,50
ISBN 3-211-83489-3

Wie stark ist die Entwicklung von Städten und Kulturlandschaften von Eigengesetzlichkeiten bestimmt? Wie form- und maßgebend sind andere zivilisatorische Zustände und Prozesse? Wenn territoriale Strukturen andere Sachverhalte bezeichnen, entwickeln sie dann eine Sprache, die ihrerseits Regeln entfaltet?

In verschiedenen thematischen Annäherungen liefert der Autor entsprechende Anregungen für das Planen und Entwerfen. Eine umsetzungsorientierte Betrachtungsweise zeigt Ansätze, wie bestehende Stadtlandschaften aktualisiert werden können, ohne die Spuren ihrer Geschichte zu verlieren und wie neue Projekte als Anstöße zu zukunftsweisenden Prozessen angelegt werden können. Das zu Grunde liegende Leitbild zielt auf ökologische Nachhaltigkeit und kulturelle Zeit- und Reichhaltigkeit.

„... Für Siedlungs- und Städtebauinteressierte ... wird diese Öffnung des Planungsbewusstseins in das Undogmatisch-Komplexe wohl zu einem Standardwerk werden."

Architektur Aktuell

SpringerWienNewYork

A-1201 Wien, Sachsenplatz 4–6, P.O. Box 89, Fax +43.1.330 24 26, e-mail: books@springer.at, Internet: www.springer.at
Birkhäuser c/o SAG, D-69126 Heidelberg, Haberstraße 7, Fax: +49.6221.345-4229, e-mail: orders@springer.de
Chronicle Books, USA, San Francisco, CA 94105, 85 Second Street, Fax +1.800.858-7787, e-mail: sales@papress.com